QUANTUM
FIELD THEORY

QUANTUM FIELD THEORY

Harald Fritzsch
Ludwig Maximilian University of Munich, Germany

NEW JERSEY • LONDON • SINGAPORE • BEIJING • SHANGHAI • HONG KONG • TAIPEI • CHENNAI • TOKYO

Published by

World Scientific Publishing Co. Pte. Ltd.

5 Toh Tuck Link, Singapore 596224

USA office: 27 Warren Street, Suite 401-402, Hackensack, NJ 07601

UK office: 57 Shelton Street, Covent Garden, London WC2H 9HE

Library of Congress Cataloging-in-Publication Data
Names: Fritzsch, Harald, 1943– author.
Title: Quantum field theory / Harald Fritzsch (Ludwig Maximilian University of Munich, Germany).
Description: Singapore ; Hackensack, NJ : World Scientific, [2017] |
Includes bibliographical references and index.
Identifiers: LCCN 2016049881| ISBN 9789813141728 (hardcover ; alk. paper) |
ISBN 9813141727 (hardcover ; alk. paper)
Subjects: LCSH: Quantum field theory.
Classification: LCC QC174.45 .F767 2017 | DDC 530.14/3--dc23
LC record available at https://lccn.loc.gov/2016049881

British Library Cataloguing-in-Publication Data
A catalogue record for this book is available from the British Library.

Desk Editor: Ng Kah Fee

Typeset by Stallion Press
Email: enquiries@stallionpress.com

Printed in Singapore

Contents

Chapter 1

Introduction

From Classical Physics to Modern Physics

The classical mechanics is valid for systems where all velocities are much less than the velocity of light. If the velocities are close to the velocity of light, the classical mechanics must be replaced by the relativistic mechanics, introduced by Albert Einstein in 1905.

In the relativity theory the speed of light is a fundamental constant:

$$c = 299\,792\,458\,\mathrm{m/sec}.$$

This implies that the flow of time depends on the reference system. Energy and mass are related to each other — mass is "frozen" energy. For a mass at rest the corresponding energy is given by the equation of Einstein: $E = mc^2$.

The *Lorentz transformations* are transformations between the space-time coordinates of one system of reference and another one. Space and time are unified to the four-dimensional space-time. The Lorentz transformations can be regarded as rotations in the four-dimensional space-time.

Electrodynamics is another important classical field of physics, which describes the behavior of electric charges and of electromagnetic fields. The theory of classical electrodynamics was introduced by James Clerk Maxwell in 1864. The basic equations of electrodynamics are the *Maxwell equations*.

Quantum Mechanics

Besides relativistic mechanics, quantum mechanics is another important branch of modern physics. Quantum physics started in 1900, when Max Planck wrote a paper on the quantization of the energy in electromagnetic processes. He introduced the new fundamental constant h.

In 1905 Albert Einstein proposed that the energy of light is quantized. Light is a collection of particles, which are called *photons*. The energy of a photon is given by the product of Planck's constant h and the frequency of the light.

In 1924 Louis de Broglie published a theory of matter waves. A particle, e.g. an electron, is at the same time also a wave, just like the case of a photon. The wave length is given by the ratio of Planck's constant and the momentum of the electron. Thus electrons and photons are similar, but photons move at the speed of light, while electrons can have any speed smaller than the speed of light.

Werner Heisenberg discovered in 1927 that in quantum physics observables cannot be precisely determined; they have a fundamental uncertainty, which is related to Planck's constant h. For example, the product of the uncertainty of the position of a particle and the uncertainty of the momentum cannot be smaller than Planck's constant:

$$\Delta x \cdot \Delta p \sim h.$$

In 1926 Erwin Schrödinger introduced the wave mechanics. He interpreted the matter waves as the wave functions of particles. The time evolution of these wave functions is described by a differential equation, the *Schrödinger equation*. He showed that his wave mechanics is equivalent to the matrix mechanics. Following this school of thought, Max Born interpreted the square of the wave function as the probability density.

In 1914 James Chadwick discovered that energy and momentum were not conserved in the beta decay of atomic nuclei. For many years this phenomenon was not understood. In 1930 Wolfgang Pauli suggested that in the beta decay not only an electron was emitted, but also a neutral particle, which could not be observed.

The energy and the momentum of this particle, later called the neutrino, would be the observed missing energy and momentum. Pauli assumed that neutrinos could never be observed directly. But in 1956 they were discovered by Clyde Cowan and Frederick Reines while investigating the neutrino emission of a big reactor in South Carolina.

Quantum Electrodynamics

In 1929 Werner Heisenberg and Wolfgang Pauli went one step further from quantum physics and quantized the electromagnetic field, which started the development of quantum electrodynamics (QED). In this theory the electromagnetic forces are generated by the exchange of photons.

The Schrödinger equation cannot be used to describe relativistic effects, since it contains the second derivatives of the three-dimensional space, but only the first derivative of time. A relativistic field equation was introduced by Paul Dirac in 1928. In the *Dirac equation* only the first derivatives with respect to space and time appear. Dirac noticed that his equation was only consistent if the electron has an antiparticle, later called the positron. The positron was discovered in 1932 in the cosmic rays.

In quantum electrodynamics there is a serious problem. An electron can emit virtual photons, but can also absorb them afterwards. If one calculates this process, one obtains a divergent result — the charge and also the mass of an electron are infinite. This problem was solved by Julian Schwinger, Richard Feynman and Freeman Dyson.

An electron without the electromagnetic interaction has a "naked mass" and a "naked charge." If the electromagnetic interaction is introduced, the contributions to the mass and to the charge are infinite. If one assumes that the naked mass and the naked charge are also infinite, but with a negative sign, the sum of the naked mass plus the corrections should be equal to the observed mass of the electron. The same can be done for the charge. Then all infinities have disappeared, due to the *renormalization* of the mass and the charge.

In quantum electrodynamics many quantities can be calculated using this renormalization technique, for example the magnetic moment of the electron. The calculated values agree with the observed values — the difference between the observed values and the calculated values is less than 10^{-8}!

Gauge Theories

In quantum electrodynamics the electron interacts with the photon. Two fields are present, the Dirac field for the electron and the vector field for the photon. If the phase of the Dirac field is changed in such a way that the change depends on space and time and if the vector field is changed by adding the space-time derivative of the phase of the Dirac field, then nothing changes. This symmetry is called a *gauge symmetry.* It was discovered in 1918 by Hermann Weyl.

Theories of this type are called *gauge theories.* The associated gauge group is the group of phase transformations, the group U(1). Thus quantum electrodynamics is a gauge theory with the gauge group U(1) — the photon is a *gauge boson.*

Wolfgang Pauli studied in 1953 a gauge theory with the gauge group SU(2). In this theory the gauge bosons are a triplet of the gauge group, thus there would be three gauge bosons without a mass. But in nature such gauge bosons do not exist, unless they have a very large mass. Pauli did not know how to introduce a mass for the gauge bosons. Thus he did not publish his idea.

But in 1954 the SU(2) gauge theory was published by Chen Ning Yang and Robert Mills, who worked in Princeton at the Institute for Advanced Study. However Yang and Mills also did not know how to introduce a mass for the gauge bosons.

After 1960 Sheldon Glashow, Abdus Salam and Steven Weinberg unified quantum electrodynamics and the theory of the weak interactions. They constructed a theory of the electroweak interactions based on the gauge group SU(2)×U(1). In this theory the weak forces are generated by the exchange of very massive gauge bosons. The theory has four gauge bosons: three massive bosons, which mediate the weak interactions, and the photon.

The masses of these bosons were generated by a spontaneous symmetry breaking. This mechanism was introduced in 1964 by Robert Brout, Francois Englert and Peter Higgs. In 1971 it was shown by Gerard 't Hooft and Martinus Veltman, that a gauge theory is renormalizable, if the masses of the gauge bosons are introduced by a spontaneous symmetry breaking.

Quantum Chromodynamics

The atomic nuclei are bound states of protons and neutrons. But these nucleons are not elementary, but consist of the quarks. A proton is a bound state of three quarks. Two different quarks are needed to describe all atomic nuclei, the up quarks and the down quarks. The electric charge of the up quark is (+2/3)e, the charge of the down quark is (–1/3)e — the electric charge of the proton is (+e).

The interactions among the quarks are described by the theory of quantum chromodynamics (QCD), introduced by Murray Gell-Mann and the author in 1972. The forces among the quarks are generated by the exchange of the gauge bosons of QCD, the *gluons.*

The quarks and gluons do not exist as free particles — they are permanently confined in the hadrons. The quarks can be observed indirectly in the scattering of electrons and atomic nuclei. Such experiments were started in 1967 at the Stanford Linear Accelerator Center in California.

Standard Model of Particle Physics

In nature there exist not only the up and down quarks, but four other quarks as well, the *strange* quark, the *charm* quark, the *bottom* quark and the *top* quark. The stable matter in the universe consists only of the up and down quarks. The other four quarks are inside unstable hadrons.

Besides the electron and its neutrino, there are also four other similar particles, the muon and the tau-lepton (or tauon), plus the two associated neutrinos. These six particles are called *leptons.*

Since there are six quarks and six leptons, it is possible to arrange them into three families. Each family consists of two quarks and two leptons. The electron, the electron neutrino, the up quark and the down quark form the first family; the muon, the muon neutrino, the charm quark and the strange quark define the second family. Lastly the third family consists of the tauon, the tau neutrino, the top quark and the bottom quark. With these three families of fundamental particles and the fundamental interactions (electroweak theory and quantum chromodynamics) put together, we have the *Standard Model of particle physics.*

It is not known, how the masses of the quarks and the leptons are generated and why the masses are quite different. The mass of the electron is only 0.5 MeV, but the mass of the top quark is very large, about 174 000 MeV.

Also we do not know, whether the Standard Model is an exact description of nature or only a good approximation. Most physicists think that at very high energies new interactions and new particles will play an important role. There are indications that there is physics beyond the Standard Model. For example, now it is known that neutrinos must have a small mass — in the Standard Model the neutrinos do not have any mass.

Beyond the Standard Model

About 80% of the matter in the universe is not the nuclear matter inside the stars, but dark matter. However, it remains unknown whether the dark matter consists of yet unknown neutral stable particles. Since 2012 one has been searching for physics beyond the Standard Model with the Large Hadron Collider (LHC) at CERN, thus far without any success.

Many physicists assume that at very high energies the electroweak theory and quantum chromodynamics are unified. In a *Grand Unified Theory* the electroweak gauge group and the color group of QCD are subgroups of a larger group. An interesting possibility is the group SO(10), introduced in 1975 by Peter Minkowski and the author (see Chapter 16).

In a Grand Unified Theory both leptons and quarks are described together in certain representations of the gauge group. Thus the proton is not stable, but can decay, e.g. into a positron and a neutral pion. One has searched for the proton decay with large detectors, e.g. with the detector Kamiokande near Kamioka in Japan, thus far without success. According to these experiments the lifetime of a proton must be larger than about 10^{32} years.

It is possible that the leptons and quarks are not elementary point-like particles, but small one-dimensional objects, the *strings*. A unification of all interactions, including gravity, might be achieved in a string theory.

Chapter 2

Mechanics

Classical Mechanics

The movement of a point mass is determined by the principle of least action. The Lagrange function of a point mass is given by the difference of the kinetic energy T and the potential energy U, which are functions of the generalized coordinates and velocities:

$$L = T(q, \dot{q}) - U(q, \dot{q}).$$

Often one also considers the total energy of the system, the Hamilton function, which is the sum of T and U:

$$H = T(q, \dot{q}) + U(q, \dot{q}).$$

The action S is given by the time integral of the Lagrange function:

$$S = \int_t^T L dt.$$

The trajectory, realized in nature, is the one for which the action is the minimum. In this case the action does not change if the trajectory is varied infinitesimally:

$$q \Rightarrow q + \delta q, \quad \dot{q} \Rightarrow \dot{q} + \delta \dot{q},$$
$$\delta S = 0.$$

The change of the action is in general given by the following integral:

$$\delta S = \int_{t_1}^{t_2} dt \left(\frac{\partial L}{\partial q} \delta q + \frac{\partial L}{\partial \dot{q}} \delta \dot{q} \right).$$

The second term can also be written as:

$$\int_{t_1}^{t_2} dt \left(\frac{\partial L}{\partial \dot{q}} \delta \dot{q} \right) = \left[\frac{\partial L}{\partial \dot{q}} \delta q \right]_{t_1}^{t_2} - \int_{t_1}^{t_2} dt \left(\delta q \frac{d}{dt} \frac{\partial L}{\partial \dot{q}} \right).$$

The variation of the coordinates at the beginning and at the end vanishes:

$$\rightarrow \int_{t_1}^{t_2} dt \left(-\frac{d}{dt} \frac{\partial L}{\partial \dot{q}} + \frac{\partial L}{\partial q} \right) \cdot \delta q = \delta S = 0.$$

Since the variation is arbitrary, one obtains the Lagrange equation:

$$\frac{d}{dt} \frac{\partial L}{\partial \dot{q}} = \frac{\partial L}{\partial q}.$$

If the system is described by several coordinates, then there is such an equation for each coordinate:

$$\frac{d}{dt} \frac{\partial L}{\partial \dot{q}_i} = \frac{\partial L}{\partial q_i}.$$

The energy of a system is given by the Lagrange function:

$$E = \sum_i \frac{\partial L}{\partial \dot{q}_i} \dot{q}_i - L.$$

For illustrative purpose we consider two examples:

a) Free point mass

$$L = \frac{m}{2} (\dot{q})^2,$$

$$m\ddot{q} = 0.$$

The solution is a uniform motion on a straight line.

b) Harmonic oscillator

Lagrange and Hamilton function:

$$L = \frac{m}{2}(\dot{q})^2 - \frac{1}{2}k \cdot q^2,$$

$$H = \frac{p^2}{2m} + \frac{1}{2}k \cdot q^2.$$

From the Lagrange function one derives the equation of motion:

$$\ddot{q} = -\omega^2 q,$$

$$\omega = \sqrt{\frac{k}{m}},$$

$$\rightarrow q(t) = A \sin(\omega t + \phi).$$

The amplitude A and the phase ϕ are arbitrary — they depend on the initial conditions.

Physical systems do not change, when the coordinate system is changed, e.g. if it is rotated or if it moves in comparison to the previous coordinate system. Such changes are called *Galilei transformations.*

As an example we consider two coordinate systems K and K′. The system K is described by the Cartesian coordinates $(x,\ y,\ z)$. The new system K′ is rotated compared to K. The transformation is given by an orthogonal matrix:

$$\begin{pmatrix} x' \\ y' \\ z' \end{pmatrix} = \begin{pmatrix} r_{11} & r_{12} & r_{13} \\ r_{21} & r_{22} & r_{23} \\ r_{31} & r_{32} & r_{33} \end{pmatrix} \begin{pmatrix} x \\ y \\ z \end{pmatrix}.$$

As a second example we consider the transition from a coordinate system K at rest to a system K′, which is moving in the direction of the x-coordinate. At the time $t = 0$ the two origins coincide. We find at the time t:

$$x' = x - vt, \quad y' = y, \quad z' = z.$$

Relativistic Mechanics

In the theory of relativity the speed of light is a constant, independent of any coordinate system: $c = 299\,792\,458\,\mathrm{m/sec}$. The speed of a body is always less than the speed of light.

The transformation from a system at rest to a moving system is a Lorentz transformation, i.e. a hyperbolic rotation of the space-time. The space-time has four dimensions: three spatial dimensions and one temporal dimension. A four-vector is given by four numbers — three for the space, one for the time:

$$x^{\mu} = (ct, x, y, z).$$

A point in the space-time is called an event. The movement of a body is described in the space-time by a world line, which is a sequence of events. The world line of a body at rest is a straight line: the space coordinates does not change, but the time changes (Fig. 2.1).

In a three-dimensional space the square of the distance between the point A and the point B is given by the sum of the squares of

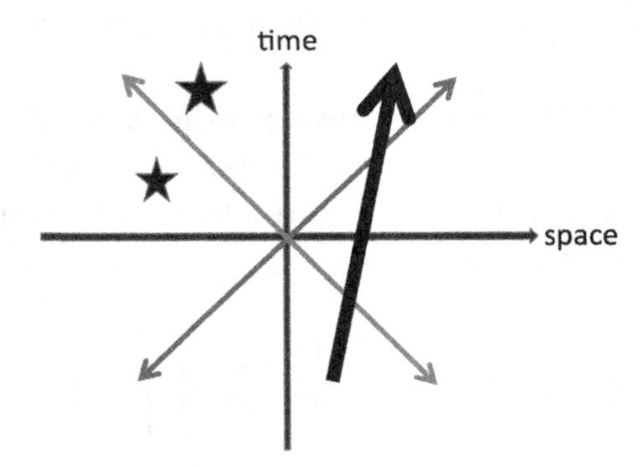

Fig. 2.1. The two-dimensional space-time. The light cone is described by two red lines. The world line of a massive particle is shown, which is moving uniformly. Also two events are shown — one is space-like with respect to the origin, the other is time-like.

the differences of the coordinates:

$$A : (X, Y, Z),$$
$$B : (x, y, z),$$
$$l^2 = (X - x)^2 + (Y - y)^2 + (Z - z)^2.$$

The distance l does not change under rotation.

Often the distance can be calculated by the *metric* of the three-dimensional space, which is given by the metric tensor:

$$g_{ik} = \begin{pmatrix} 1 & 0 & 0 \\ 0 & 1 & 0 \\ 0 & 0 & 1 \end{pmatrix},$$

$$dl^2 = g_{ik} dx^i dx^k = dx^2 + dy^2 + dz^2.$$

In the four-dimensional space-time an event is described by a four-vector:

$$x^\mu = (ct, x, y, z).$$

If we consider a light signal, emitted at $t = 0$ from the origin of space, at some time after the emission, the space and the time of the signal are related by:

$$c^2t^2 = x^2 + y^2 + z^2,$$
$$\Rightarrow c^2t^2 - x^2 - y^2 - z^2 = 0.$$

This relation must also be valid in another reference system with space-time coordinates (cT, X, Y, Z):

$$c^2t^2 - x^2 - y^2 - z^2 = c^2T^2 - X^2 - Y^2 - Z^2 = 0.$$

A Lorentz transformation can be used to describe the transition between two systems. It is given by a pseudo-orthogonal 4×4-matrix:

$$\begin{pmatrix} ct' \\ x' \\ y' \\ z' \end{pmatrix} = \begin{pmatrix} l_{00} & l_{01} & l_{02} & l_{03} \\ l_{10} & l_{11} & l_{12} & l_{13} \\ l_{20} & l_{21} & l_{22} & l_{23} \\ l_{30} & l_{31} & l_{32} & l_{33} \end{pmatrix} \begin{pmatrix} ct \\ x \\ y \\ z \end{pmatrix}.$$

A Lorentz transformation can be interpreted as a hyperbolic rotation of the four-dimensional space-time. The square of the distance between two events does not change under a Lorentz transformation:

$$s^2 = c^2(T-t)^2 - (X-x)^2 - (Y-y)^2 - (Z-z).^2$$

Also the infinitesimal distance does not change:

$$ds^2 = c^2dt^2 - dx^2 - dy^2 - dz^2 = c^2dt'^2 - dx'^2 - dy'^2 - dz'^2 = ds'^2.$$

Similar to the three-dimensional case, this can be written with the help of the metric tensor:

$$ds^2 = g_{\mu\nu}dx^\mu dx^\nu.$$

The metric tensor is a diagonal matrix:

$$g_{\mu\nu} = \begin{pmatrix} 1 & 0 & 0 & 0 \\ 0 & -1 & 0 & 0 \\ 0 & 0 & -1 & 0 \\ 0 & 0 & 0 & -1 \end{pmatrix}.$$

The distance between two events can be positive (time-like) or negative (space-like). The distance between two events vanishes if the two events can be connected by a light ray. Events, which are before the origin, describe the past light cone; events, which are after the origin, the future light cone (Fig. 2.2).

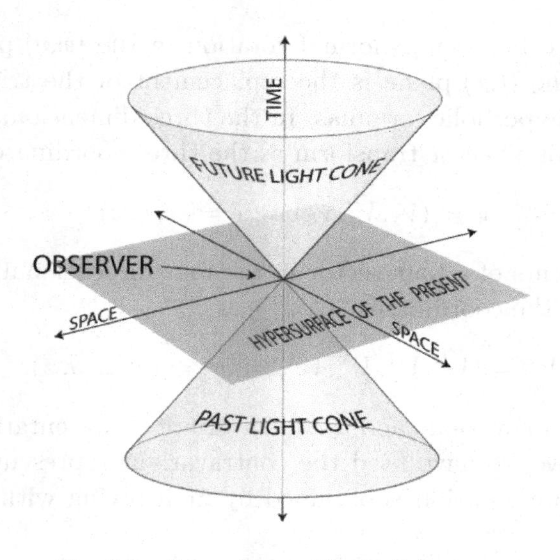

Fig. 2.2. Space-time with the light cone.

A general Lorentz transformation is given by a 4×4-matrix and a four-vector, which describes a translation:

$$x'^{\mu} = \Lambda^{\mu}_{\nu} \cdot x^{\nu} + a^{\mu}.$$

The Lorentz matrix must obey the following condition:

$$g_{\mu\nu}\Lambda^{\mu}_{\alpha}\Lambda^{\nu}_{\beta} = g_{\alpha\beta}.$$

We restrict ourselves to proper Lorentz transformations:

$$\det \Lambda = 1,$$

$$\Lambda^{0}_{0} \geq 1.$$

A transformation from a system at rest to a moving system can be considered as a hyperbolic rotation of the space and the time:

$$ct' = -x \cdot \sinh \psi + ct \cdot \cosh \psi,$$

$$x' = x \cdot \cosh \psi - ct \cdot \sinh \psi,$$

$$\tanh \psi = \frac{v}{c}.$$

The difference between a normal rotation in the (x,y)-plane and a rotation in the (t,x)-plane is the replacement of the trigonometric formulas by hyperbolic formulas. In the three-dimensional space the components of a vector transform as the three coordinates:

$$\vec{V} = (V_1, V_2, V_3) \Leftrightarrow \vec{x} = (x, y, z).$$

The components of a four-vector in the four-dimensional space-time transform as the coordinates $(ct,\ x,\ y,\ z)$:

$$V^\mu = (V^0, V^1, V^2, V^3) \Leftrightarrow x^\mu = (ct, x, y, z).$$

There is a covariant and a contravariant representation of four-vectors. Above we have used the contravariant representation. The covariant representation is obtained by multiplying with the metric tensor:

$$V_\mu = \sum_{\nu=0}^{3} g_{\mu\nu} V^\nu \equiv g_{\mu\nu} V^\nu,$$

$$V_0 = V^0, \quad V_1 = -V^1, \ldots$$

Similarly, the contravariant metric tensor is defined by the relation:

$$g^{\lambda\mu} g_{\mu\nu} = g^\lambda_\nu = \delta^\lambda_\nu,$$

$$\delta^\rho_\mu = 1 \quad (\mu = \rho),$$

$$\delta^\rho_\mu = 0 \quad (\mu \neq \rho).$$

We consider the derivative of a function $F(x)$, which is a covariant four-vector:

$$\frac{\partial F}{\partial x^\mu} = \partial_\mu F = \left(\frac{\partial F}{\partial x^0}, \frac{\partial F}{\partial x^1}, \frac{\partial F}{\partial x^2}, \frac{\partial F}{\partial x^3} \right).$$

There is also another type of derivative, which is a contravariant four-vector:

$$\frac{\partial F}{\partial x_\mu} = \partial^\mu F.$$

We can check, that the scalar product of a covariant and a contravariant four-vector is invariant under Lorentz transformations:

$$x^\mu x_\mu = x'^\mu x'_\mu = (x^0)^2 - (\vec{x})^2,$$
$$\Rightarrow \Lambda^{\lambda\mu}\Lambda_{\lambda\nu} = \delta^\mu_\nu,$$
$$V_\mu V^\mu = V_0^2 - V_x^2 - V_y^2 - V_z^2.$$

Recall that the action is the integral of the Lagrange function:

$$S = \int_t^T L dt.$$

In relativistic mechanics the action of a free point mass is given by:

$$S = -mc \int_a^b ds.$$

The Lagrange function of a free point mass and its momentum are given by

$$L = -mc^2\sqrt{1 - \frac{v^2}{c^2}},$$
$$\vec{p} = \frac{\partial L}{\partial \vec{v}} = \frac{m\vec{v}}{\sqrt{1 - \frac{v^2}{c^2}}}.$$

If the speed v is much less than the speed of light, one has:

$$\vec{p} \cong m\vec{v}.$$

In the nonrelativistic classical mechanics the momentum of a point mass is a vector, the energy is a scalar. In the relativistic mechanics energy and momentum form a four-vector:

$$p^\mu = (E/c, p_x, p_y, p_z).$$

The square of this four-vector is invariant under Lorentz transformations:

$$(p_\mu \cdot p^\mu) \cdot c^2 = E^2 - \vec{p}^2 c^2 = m^2 c^4.$$

Here m is the rest mass of the particle. If the momentum vanishes, one obtains the famous equation of Einstein:

$$E = mc^2.$$

In general one has the following relation between energy and momentum:

$$E = c\sqrt{\vec{p}^2 + m^2c^2}.$$

If the momentum p is much smaller than (mc), one obtains the kinetic energy in the nonrelativistic classical mechanics:

$$\tilde{E} = E - mc^2 \approx \frac{\vec{p}^2}{2m}$$

From the discussion above we can see that the nonrelativistic classical mechanics follows from the relativistic mechanics, if all velocities are small in comparison to the speed of light.

For a massless particle the energy is given by the momentum, multiplied with the speed of light:

$$E = c \cdot p.$$

Chapter 3

Classical Fields

Scalar Fields

A system of point masses is described by the Lagrange equations:

$$\frac{d}{dt}\frac{\partial L}{\partial \dot{q}_i} = \frac{\partial L}{\partial q_i}.$$

Now we replace the discrete points by continuous fields:

$$q_i \rightarrow \phi(x),$$

$$\dot{q}_i \rightarrow \partial_\mu \phi(x).$$

The Lagrange function is replaced by the Lagrange density, which is a function of the field and the first space and time derivatives of the field:

$$L = L(\phi, \partial_\mu \phi).$$

The action $S(G)$ is the integral of the Lagrange density over a finite volume G in the four-dimensional space-time:

$$S(G) = \int_G d^4x \cdot L(\phi, \partial_\mu \phi).$$

Now we vary the field in such a way that the variation on the surface of G vanishes:

$$\phi(x) \Rightarrow \phi(x) + \delta\phi(x),$$

$$\delta\phi(x \propto G) = 0.$$

We require that the action integral does not change under this variation:

$$\delta S(G) = \int_G d^4x \left(\frac{\partial L}{\partial \phi} \delta\phi + \frac{\partial L}{\partial(\partial_\mu \phi)} \delta(\partial_\mu \phi) \right) = 0.$$

This integral can also be written as follows:

$$\delta S(G) = \int_G \left\{ \left(\frac{\partial L}{\partial \phi} - \frac{\partial}{\partial x^\alpha} \frac{\partial L}{\partial(\partial_\alpha \phi)} \right) \delta\phi + \frac{\partial}{\partial x^\alpha} \left(\frac{\partial L}{\partial(\partial_\alpha \phi)} \delta\phi \right) \right\} \cdot d^4x.$$

The last term can be changed into a surface integral, which vanishes, since the field at large space-time coordinates vanishes. The variation of the field is arbitrary, thus we obtain the field equation, which is analogous to the Lagrange equation:

$$\frac{\partial L}{\partial \varphi} = \frac{\partial}{\partial x^\mu} \left(\frac{\partial L}{\partial(\partial_\mu \phi)} \right).$$

The momentum of the field is the derivative of the Lagrange density with respect to the time derivative of the field:

$$\pi(x) = \frac{\partial L}{\partial \dot{\phi}}.$$

As an example we consider the Lagrange density of a real massive free scalar field with mass m:

$$L = \frac{1}{2}\dot{\phi}^2 - \frac{1}{2}(\vec{\nabla}\phi)^2 - \frac{1}{2}m^2\phi^2 = \frac{1}{2}((\partial_\mu \phi)^2 - m^2\phi^2).$$

The field equation is the Klein–Gordon equation:

$$\left(\frac{\partial^2}{\partial t^2} - \vec{\nabla}^2 + m^2 \right) \phi = 0,$$

$$(\partial_\mu \partial^\mu + m^2)\phi(x) = 0.$$

Another example is the Lagrange density of an interacting field in two dimensions:

$$L = \frac{1}{2} \left\{ (\partial_t \varphi)^2 - (\partial_x \varphi)^2 \right\} + \cos\phi.$$

The field equation is the Sine-Gordon equation:

$$\frac{\partial^2\varphi}{\partial t^2} - \frac{\partial^2\varphi}{\partial x^2} = -\sin\phi.$$

We will see later that important for the spontaneous symmetry breaking is the following Lagrange density:

$$L = \frac{1}{2}(\partial^\mu\varphi \cdot \partial_\mu\varphi) - \frac{1}{2}m^2\varphi^2 - \frac{1}{4}\lambda\varphi^4.$$

Electrodynamics

The electric and magnetic fields are vector fields. They are described by a four-potential:

$$A^\mu = (A^0, \vec{A}),$$
$$A_\mu = (A_0, -\vec{A}).$$

The three space components of the potential are the vector potential, the time component is the scalar potential:

$$A^\mu = (\varphi, \vec{A}).$$

Using the four components of the potential, one can calculate the electric and magnetic field strength:

$$\vec{E} = -\frac{\partial\vec{A}}{\partial t} - \vec{\nabla}\varphi,$$

$$\vec{H} = \nabla \times \vec{A}.$$

The field strength tensor is given by the electric field strengths E and the magnetic field strengths H:

$$F^{\mu\nu} = \partial^\mu A^\nu - \partial^\nu A^\mu = -F^{\nu\mu},$$

$$F_{\alpha\beta} = g_{\alpha\mu} \cdot g_{\beta\nu} \cdot F^{\mu\nu},$$

$$(F^{\mu\nu}) = \begin{pmatrix} 0 & -E^1 & -E^2 & -E^3 \\ E^1 & 0 & -H^3 & H^2 \\ E^2 & H^3 & 0 & -H^1 \\ E^3 & -H^2 & H^1 & 0 \end{pmatrix}.$$

The electric and magnetic field strengths do not change under a gauge transformation:

$$A'_\mu = A_\mu + \frac{\partial f(x)}{\partial x^\mu}.$$

Here $f(x)$ is a scalar function of the four space-time coordinates. By a suitable gauge transformation it can be arranged, that the divergence of the four-potential vanishes ("Lorentz condition"):

$$\partial_\mu A^\mu = 0.$$

Using the definition of the field strength tensor, one can derive the field equation:

$$\partial_\rho F^{\mu\nu} + \partial_\mu F^{\nu\rho} + \partial_\nu F^{\rho\mu} = 0.$$

From this equation follows the first group of the Maxwell equations:

$$\nabla \times \vec{E} + \frac{\partial \vec{H}}{\partial t} = 0,$$

$$\vec{\nabla} \vec{H} = 0.$$

The interaction of a particle with an external electromagnetic field is given by the following action:

$$S = \int_a^b (-m \cdot ds + e \cdot A_\mu dx^\mu)$$
$$= \int_a^b (-m \cdot ds + e \cdot \vec{A} \cdot d\vec{r} - e \cdot \varphi \cdot dt).$$

This integral can also be written as a time integral:

$$S = \int_a^b (-m \cdot c^2 \sqrt{1 - v^2} + e \cdot \vec{A} \cdot \vec{v} - e \cdot \varphi) \cdot dt = \int_a^b L \cdot dt,$$

$$\vec{v} = \frac{d\vec{r}}{dt},$$

$$L = (-m \cdot c^2 \sqrt{1 - v^2} + e \cdot \vec{A} \cdot \vec{v} - e \cdot \varphi).$$

The generalized momentum is obtained by differentiating L with respect to time. It is the sum of the normal momentum and of the vector potential multiplied with e:

$$\vec{P} = \vec{p} + e \cdot \vec{A}.$$

If the velocity is small, one finds the Lagrange density:

$$L \cong \frac{1}{2} m \cdot v^2 + e \cdot \vec{A} \cdot \vec{v} - e \cdot \varphi.$$

Here the equation of motion of a charge in an electromagnetic field is given by:

$$\frac{d\vec{p}}{dt} = e \cdot \vec{E} + e \cdot (\vec{v} \times \vec{H}).$$

The first term is the force generated by the electric field. The second term is the "Lorentz force", which is proportional to the velocity of the charge and perpendicular to the magnetic field.

Let us now consider the source of electromagnetic fields: the electric charges. A charge at rest is described by the charge density. The integral of the charge density gives the total charge. The charge density of a point particle, e.g. of an electron, which is at the origin of space coordinates, is given by a three-dimensional delta function:

$$\rho(\vec{x}) = \delta^3(\vec{x}).$$

On the other hand a moving charge is described by the current j:

$$\vec{j} = \rho \cdot \vec{v}.$$

In four-dimensional space-time the charge density and the current are the components of a four-dimensional current:

$$j^\mu = (j^0, \vec{j}) = (\rho, \vec{j}),$$

$$j_\mu = (j^0, -\vec{j}),$$

$$j^0 = j_0 = \rho.$$

This current is conserved:

$$\partial_\mu j^\mu = 0,$$

$$\Rightarrow \frac{\partial \rho}{\partial t} + \vec{\nabla} \cdot \vec{j} = 0.$$

The divergence of the electric field is proportional to the charge density:

$$\vec{\nabla}\vec{E} = \rho.$$

From this relation and from an analogous relation for the magnetic field strength one can calculate the divergence of the field strength tensor:

$$\partial_\nu F^{\mu\nu} = j^\nu.$$

This gives the second group of Maxwell equations:

$$\vec{\nabla} \times \vec{H} - \frac{\partial \vec{E}}{\partial t} = \vec{j},$$

$$\vec{\nabla}\vec{E} = \rho.$$

The action of a system of particles in an electromagnetic field is the sum of the actions of the free particles, the action of the interaction of the particles with the field and the action of the electromagnetic field:

$$S = S_T + S_{WW} + S_F,$$

$$S_T = -\sum_i m_i \int ds,$$

$$S_{WW} = -\sum_i e_i \int A_\mu dx^\mu,$$

$$S_F = -\frac{1}{16\pi} \int F_{\mu\nu} \cdot F^{\mu\nu} \cdot dt \cdot d^3x = \frac{1}{8\pi} \int (E^2 - H^2) \cdot dt \cdot d^3x.$$

Finally we mention the energy-momentum tensor of the electromagnetic field:

$$T^{\mu\nu} = \frac{1}{4\pi} \left(-F^{\mu\rho} \cdot F^\nu_\rho + \frac{1}{4} g^{\mu\nu} \cdot F_{\rho\sigma} \cdot F^{\rho\sigma} \right).$$

Chapter 4

Quantum Theory

Nonrelativistic Quantum Mechanics

In quantum physics a system is described by a wave function. If the system consists only of one particle, the absolute square of the wave function describes the probability to find the particle at any specific position.

The wave function is a vector in a Hilbert space. Here we describe some properties of a Hilbert space:

1) The sum of two wave functions, multiplied with complex numbers, describes also a wave function, i.e. a vector in the same Hilbert space:

$$\Phi, \Psi \Rightarrow \alpha \cdot \Phi + \beta \cdot \Psi.$$

2) For a pair of vectors one can define a scalar product, given by a complex number:

$$(\Phi, \Psi) = (\Psi, \Phi)^* = \alpha,$$
$$(\Phi, \Phi) \geq 0.$$

3) A unit vector in the Hilbert space is a wave function with the property $(\Phi, \Phi) = 1$.

An example for a Hilbert space are the complex functions. The scalar product of two functions is given by the following integral:

$$(\Psi, \Phi) = \int_{-\infty}^{+\infty} (\Psi(x))^* \Phi(x) dx.$$

In a Hilbert space we can define operators to act on the vectors. A linear operator in the Hilbert space is a linear transformation:

$$\Phi \Rightarrow A\Phi.$$

For a Hilbert space with a finite number of dimensions this transformation is described by a matrix.

For each operator A we can define an adjoint operator $A^\dagger$ by the relation:

$$(\Phi, A^\dagger \Psi) = (A\Phi, \Psi) = (\Psi, A\Phi)^*.$$

An operator is said to be Hermitian, if the operator and the adjoint operator are identical:

$$A^\dagger = A.$$

On the other hand an operator U is said to be unitary, if the following condition for arbitrary wave functions is satisfied:

$$(U\Psi, U\Phi) = (\Psi, \Phi).$$

In a finite-dimensional Hilbert space a unitary operator is described by a unitary matrix. The exponential of a Hermitian operator H is a unitary operator U:

$$\exp(iH) = U.$$

A wave function is called an eigenvector or eigenstate of an operator, if the operator applied to the wave function reproduces the wave function multiplied with a scalar eigenvalue E:

$$A\Psi = E\Psi.$$

A Hermitian operator has real eigenvalues. Eigenvectors with different eigenvalues are orthogonal to each other:

$$\begin{aligned}
A\Psi_1 &= E_1\Psi_1, \\
A\Psi_2 &= E_2\Psi_2, \\
(E_2 - E_1)(\Psi_2, \Psi_1) &= (A\Psi_2, \Psi_1) - (\Psi_2, A\Psi_1) = 0, \\
&\Rightarrow (\Psi_2, \Psi_1) = 0.
\end{aligned}$$

The observables of a physical system are described by operators. Examples are the position of a particle, its momentum, its angular momentum and its energy. The momentum operator is proportional to the space derivative of the wave function:

$$\vec{p} = \frac{\hbar}{i}\vec{\nabla}.$$

In a quantum system the time evolution of a wave function can be calculated, using the Schrödinger equation:

$$i\hbar\frac{\partial}{\partial t}\Psi = H\Psi.$$

As an example we consider the Hamilton operator of a particle in an external field, which is described by a potential U:

$$H = \frac{\vec{p}^2}{2m} + U(\vec{r}).$$

Then the corresponding Schrödinger equation is:

$$i\hbar\frac{\partial\Psi}{\partial t} = -\frac{\hbar^2}{2m}\Delta\Psi + U(\vec{r})\cdot\Psi.$$

If a system is described by an eigenstate of the operator, the energy is given by the eigenvalue E :

$$H\Psi = E\Psi.$$

We consider the commutation relations of the coordinate and the momentum:

$$[q, p] = i\hbar.$$

Since the coordinate q and the associated momentum p do not commute, it is impossible to measure them exactly. Both the coordinate and the momentum have an uncertainty. The commutation relation implies that the product of these two uncertainties is given by Planck's constant h ("uncertainty relation"):

$$\Delta q \cdot \Delta p \geq \frac{\hbar}{2}.$$

Now we consider the Hamilton operator of a one-dimensional harmonic oscillator:

$$H = \frac{p^2}{2m} + \frac{1}{2}m\omega^2 q^2.$$

We introduce a creation operator and an annihilation operator:

$$a^\dagger = \sqrt{\frac{m\omega}{2}}\left(q - \frac{i}{m\omega}p\right),$$

$$a = \sqrt{\frac{m\omega}{2}}\left(q + \frac{i}{m\omega}p\right).$$

The commutation relations of these operators are:

$$[a, a^\dagger] = 1,$$

$$[a, a] = [a^\dagger, a^\dagger] = 0.$$

The Hamilton operator can now be written as follows:

$$H = \frac{1}{2}\cdot\left(a^\dagger a + \frac{1}{2}\right)\cdot\omega.$$

Here are the commutation relations of H with the new operators:

$$[H, a] = -\omega \cdot a,$$

$$[H, a^\dagger] = +\omega \cdot a^\dagger.$$

We define a particle number operator N:

$$N = a^\dagger a,$$

$$H = \frac{1}{2}\left(N + \frac{1}{2}\right)\cdot\omega.$$

For an eigenstate of H we have:

$$H|n\rangle = \omega \cdot \left(n + \frac{1}{2}\right) \left|n\right\rangle,$$
$$E_n = \omega \left(n + \frac{1}{2}\right).$$

The number n is an integer number. The energy of the ground state is not zero, but has a finite value.

An eigenstate of H is also an eigenstate of N:

$$N|n\rangle = n|n\rangle.$$

Here are the commutation relations for the operator N:

$$[N, a] = -a,$$
$$[N, a^\dagger] = a^\dagger.$$

If the annihilation operator is applied to an eigenstate, the number n decreases by one. If the creation operator is applied, the number n increases:

$$a|n\rangle = \sqrt{n}|n-1\rangle,$$
$$a^\dagger|n\rangle = \sqrt{n+1}|n+1\rangle.$$

Since n is a positive integer number, there is a lowest state, the ground state, often also called the "vacuum state". If the annihilation operator is applied to the ground state, one obtains zero:

$$a|0\rangle = 0.$$

The state $|1\rangle$ is obtained by applying the creation operator on the ground state:

$$|1\rangle = a^\dagger|0\rangle.$$

In a similar way one obtains the other states:

$$|n\rangle = \frac{(a^\dagger)^n}{\sqrt{n!}}|0\rangle.$$

Schrödinger Picture, Heisenberg Picture and Interaction Picture

These three pictures are three different descriptions of the time evolution of a quantum system. In the Schrödinger picture the time evolution is given by the time dependence of the wave function:

$$i\frac{\partial}{\partial t}|\Psi(t)\rangle_S = H|\Psi(t)\rangle_S.$$

The time evolution of the wave function is described by a unitary operator U:

$$|\Psi(t)\rangle_S = U(t,t_0)|\Psi(t_0)\rangle_S,$$
$$U(t,t_o) = e^{-iH(t-t_o)}.$$

In the Heisenberg picture the wave functions do not depend on time. The time evolution is described by time-dependent operators. The wave function in the Heisenberg picture is given by a unitary transformation of the corresponding wave function in the Schrödinger picture:

$$|\Psi(t)\rangle_H = U(t)^\dagger|\Psi(t)\rangle_S = |\Psi(t_0)\rangle_S.$$

Here we give the connection between operators in the Heisenberg picture and in the Schrödinger picture:

$$O^H(t) = U(t)^\dagger O^S U(t).$$

The time evolution of a Heisenberg operator is given by the commutator with the Hamilton operator:

$$i\frac{d}{dt}O^H(t) = [O^H(t), H].$$

Besides the Schrödinger picture and the Heisenberg picture one uses also the interaction picture. Often the Hamilton operator is the sum of two terms, the free Hamilton operator and an operator for describing the interaction:

$$H = H_0 + H_{int}.$$

We can consider the unitary transformation:

$$U_0(t, t_o) = e^{-iH_0(t-t_o)}.$$

A wave function in the interaction picture is given by the following transformation:

$$|\Psi(t)\rangle_{\text{int}} = U_0^\dagger |\Psi(t)\rangle_S.$$

The time derivative of this wave function is given by the interaction Hamiltonian:

$$i\frac{\partial}{\partial t}|\Psi(t)\rangle_{\text{int}} = H_{\text{int}}|\Psi(t)\rangle_{\text{int}}.$$

An operator in the interaction picture is given by the unitary transformation:

$$O^{\text{int}}(t) = e^{iH_0 \cdot t} O^S e^{-iH_0 \cdot t}.$$

The time evolution for an operator in the interaction picture is determined by the commutator with the free Hamilton operator:

$$i\frac{d}{dt}O^{int}(t) = [O^{\text{int}}(t), H_0].$$

Relativistic Quantum Mechanics

A state vector in the Hilbert space is transformed to a new state vector by a Lorentz transformation. This transformation is a unitary transformation:

$$|\Psi\rangle \Rightarrow U(a, \Lambda)|\Psi\rangle.$$

The generators of the translations in the four-dimensional space-time are the four momenta:

$$U(a) = \exp(iP^\mu a_\mu).$$

The proper Lorentz transformations are described by six generators:

$$U(\Lambda) = \exp\left(\frac{i}{2}\alpha_{\mu\nu} M^{\mu\nu}\right),$$

$$\alpha_{\mu\nu} = -\alpha_{\nu\mu}.$$

Here are the commutation relations of all generators:

$$[P^\mu, P^\nu] = 0,$$

$$[M^{\mu\nu}, P^\sigma] = -i(P^\mu g^{\nu\sigma} - P^\nu g^{\mu\sigma}),$$

$$[M^{\mu\nu}, M^{\rho\sigma}] = i(M^{\mu\rho} g^{\nu\sigma} + M^{\nu\sigma} g^{\mu\rho} - M^{\nu\rho} g^{\mu\sigma} - M^{\mu\sigma} g^{\nu\rho}),$$

Now we define the operator M^2:

$$M^2 = P_\mu P^\mu.$$

This operator commutes with all 10 generators:

$$[M^2, P^\mu] = 0,$$

$$[M^2, M^{\mu\nu}] = 0.$$

The operator M^2 is the mass operator. If it is applied on a mass eigenstate, one obtains the square of the mass of the state.

The Schrödinger equation is a nonrelativistic equation. On the left is the derivative of the wave function with respect to the time, on the right are the second derivatives with respect to the three space coordinates.

In the theory of special relativity the energy and the three momenta are the four components of a four-vector:

$$p^\mu = (p^0, p^1, p^2, p^3) = (E, p_x, p_y, p_z).$$

The scalar product of this four-vector with itself gives the square of the mass:

$$p^\mu p_\mu = E^2 - \vec{p}^2 = m^2.$$

If one replaces the energy by the time derivative and the three momenta by the three space derivatives, one obtains the Klein–Gordon equation:

$$E \rightarrow i\hbar \frac{\partial}{\partial t}, \quad \vec{p} \rightarrow -i\hbar \vec{\nabla},$$

$$\left[\partial_\mu \partial^\mu + \left(\frac{mc}{\hbar}\right)^2\right] \varphi(x) = 0.$$

If a wave function is described by the Schrödinger equation, one can obtain a positive definite probability density. For the Klein–Gordon equation this is not possible, since it contains a second derivative with respect to time. This is a problem, and the Klein–Gordon equation is not a relativistic generalization of the Schrödinger equation. But we shall use it later for the description of the dynamics of scalar fields.

The particles have an intrinsic angular momentum, which is called the "spin". All particles have either integer spin or half-integer spin. The spin statistics theorem states that the wave function of a system of identical integer-spin particles has the same value when the positions of any two particles are swapped. These particles with wave functions symmetric under interchange are called bosons.

The wave function of a system of identical half-integer spin particles changes sign when two particles are swapped. These particles with wave functions antisymmetric under interchange are called fermions. Bosons obey the Bose–Einstein statistics, while fermions obey the Fermi–Dirac statistics.

Chapter 5

Group Theory

Group theory is a field of mathematics that has become important for physics, especially for the understanding of symmetries. A group is a set of elements (a, b, c, ...), which can be multiplied, following a specific set of rules:

(a) The multiplication of two elements a and b gives another element c:

$$a \in G, \quad b \in G \quad \Rightarrow a \times b = c \in G.$$

(b) The multiplication is associative:

$$(a \times b) \times c = a \times (b \times c).$$

(c) The group has a unit element e:

$$a \times e = e \times a = a.$$

(d) For each element there is an inverse element. The multiplication of an element with the inverse element gives the unit element:

$$a \times (a^{-1}) = (a^{-1}) \times a = e.$$

If the group is an Abelian group, the multiplication is commutative:

$$a \times b = b \times a.$$

An example is the group of rational numbers without zero.

Lie Groups

Important in particular for quantum physics are the Lie groups. The elements of a Lie group are functions of a set of parameters. An element can be described as a function of these parameters:

$$G = G(\theta_1, \theta_2, \dots, \theta_n).$$

The unit element is the element for which all parameters are zero:

$$e = G(0, 0, \dots, 0).$$

The derivatives of the unit element define the generators of the group:

$$T_i = \left. \frac{\partial G}{\partial \theta_i} \right|_{\theta_i = 0}.$$

The commutation relations of the generators define the Lie algebra of the group:

$$[T_i, T_j] = i f_{ijk} T_k.$$

The constants f_{ijk} are the structure constants of the Lie algebra. They are antisymmetric and obey the Jacobi identity:

$$[T_a, [T_b, T_c]] + [T_b, [T_c, T_a]] + [T_c, [T_a, T_b]] = 0,$$

$$\rightarrow f_{ade} f_{bcd} + f_{bde} f_{cad} + f_{cde} f_{abd} = 0.$$

The sum of the squares of all generators T does not change under a group transformation, since it commutes with all generators:

$$T^2 = \sum_a T_a T_a,$$

$$[T_a, T^2] = 0.$$

If a Lie group does not have any subgroup, it is a simple group. There are three families of such groups: Unitary groups $\mathrm{U}(n)$, orthogonal groups $\mathrm{O}(n)$ and symplectic groups $\mathrm{Sp}(n)$. Besides these three families there are five exceptional groups:

$$G_2(14), \quad F_4(52), \quad E_6(78), \quad E_7(133), \quad E_8(248).$$

The number of generators of these groups are given in the parentheses. The group E(8) is interesting: The smallest representation of E(8) is the adjoint representation, which has 248 elements. But it seems that the exceptional groups and the symplectic groups are not relevant for physics.

Orthogonal Groups

The group SO(n) is the group of n-dimensional orthogonal matrices. In an n-dimensional space one has $n(n-1)/2$ independent rotations — the group SO(n) has therefore $n(n-1)/2$ generators.

The simplest orthogonal group is the group SO(2), which describes the rotations in a two-dimensional space. It has one generator and is the group of two-dimensional orthogonal matrices, which are determined by an angle:

$$g = \begin{pmatrix} \cos\varphi & \sin\varphi \\ -\sin\varphi & \cos\varphi \end{pmatrix}.$$

The group SO(3) is the group of rotations in a three-dimensional space. These rotations can be described by three angles, e.g. the three Euler angles. The Lie algebra of SO(3) is identical to the algebra of the generators of the angular momentum in the classical mechanics:

$$[J_i, J_j] = i\varepsilon_{ijk} J_k.$$

The Casimir operator J^2 is the sum of the squares of the three generators:

$$J^2 = J_1^2 + J_2^2 + J_3^2,$$

$$[J_i, J^2] = 0.$$

Later we shall discuss the SO(10) — this group might be relevant for physics, since it can describe the unification of the strong and electroweak interactions, as described in Chapter 16.

Unitary Groups

The unitary groups U(n) are the groups of n-dimensional unitary matrices. The groups SU(n) are subgroups of U(n) with the constraint that the determinant is equal to 1.

The generators of SU(n) are n-dimensional Hermitian matrices with a vanishing trace. There are $(n^2 - 1)$ such matrices. The rank of a group is the number of generators, which can be written as diagonal matrices. The rank of SU(2) is 1, the rank of SU(3) is 2, etc.

The simplest unitary group is the Abelian group U(1). The group elements are all complex numbers with the absolute value (+1):

$$U = \exp(-i\theta).$$

The group SU(2) is the group of unitary two-dimensional matrices with the constraint $\det(U) = 1$. This group describes in particular the symmetry of isospin in nuclear and particle physics.

The groups SU(2) and SO(3) are isomorphic. The Lie algebra of SU(2) is identical to the Lie algebra of SO(3), the algebra of the angular momentum operators:

$$[J_i, J_j] = i\varepsilon_{ijk} J_k.$$

Since an SU(2) transformation can also be interpreted as an SO(3) transformation, it is described by three real parameters, which correspond to the three rotation angles in a three-dimensional space. Here we denote for rotations around the three coordinate axes the corresponding SU(2) transformations:

$$U = \begin{pmatrix} \cos\alpha/2 & i\sin\alpha/2 \\ i\sin\alpha/2 & \cos\alpha/2 \end{pmatrix},$$

$$U = \begin{pmatrix} \cos\beta/2 & i\sin\beta/2 \\ i\sin\beta/2 & \cos\beta/2 \end{pmatrix},$$

$$U = \begin{pmatrix} e^{i\gamma/2} & 0 \\ 0 & e^{-i\gamma/2} \end{pmatrix}.$$

As mentioned before, the groups SO(3) and SU(2) have the same structure — they are isomorphic:

$$SO(3) \cong SU(2).$$

The group SO(4) is isomorphic to a product of two SU(2) groups:

$$SO(4) \cong SU(2) \otimes SU(2).$$

The group SO(6) is isomorphic to SU(4):

$$SO(6) \cong SU(4).$$

These relations will be important later, when we discuss the group SO(10), which describes the Grand Unification of the electroweak and strong interactions.

Chapter 6

Free Scalar Fields

Real Scalar Fields

A quantized scalar field describes scalar particles, i.e. particles without spin. First we discuss the Lagrange density of a real scalar field, which describes neutral particles:

$$L = \frac{1}{2}\dot{\varphi}^2 - \frac{1}{2}(\vec{\nabla}\varphi)^2 - \frac{1}{2}m^2\varphi^2 = \frac{1}{2}\partial_\mu\varphi \cdot \partial^\mu\varphi - \frac{1}{2}m^2\varphi^2.$$

This Lagrange density is similar to the Lagrange function of a one-dimensional harmonic oscillator:

$$L = \frac{m}{2}(\dot{q})^2 - \frac{1}{2}k \cdot q^2.$$

The action of a field is the integral of the Lagrange density over a finite volume of the space-time:

$$S = \frac{1}{2}\int_G d^4x(\partial_\mu\varphi \cdot \partial^\mu\varphi - m^2\varphi^2).$$

The corresponding Lagrange field equation is the Klein–Gordon equation:

$$\left(\frac{\partial^2}{\partial t^2} - \vec{\nabla}^2 + m^2\right)\varphi(t, \vec{x}) = 0,$$

$$(\partial_\mu\partial^\mu + m^2)\varphi(x) = 0.$$

The energy, the momentum and the mass of the associated scalar particle are related:

$$E^2 = p^2 + m^2,$$

$$E = \pm\sqrt{p^2 + m^2}.$$

Thus the energy can be positive or negative. In the field theory the states with negative energy are interpreted as antiparticles with positive energy. For a real scalar field particles and antiparticles are identical.

The momentum density of a field is given by the derivative of the Lagrange density with respect to time:

$$\pi(x) = \frac{\partial L}{\partial \dot{\varphi}(x)} = \dot{\varphi}(x).$$

Here are the Hamilton density, the energy and the momentum of the field:

$$H(x) = \pi\dot{\varphi} - L = \frac{1}{2}(\pi^2 + (\vec{\nabla}\varphi)^2 + m^2\varphi^2),$$

$$E = P^0 = \int d^3x \cdot H(x),$$

$$\vec{P} = -\int \pi \cdot \vec{\nabla}\varphi \cdot d^3x.$$

In the nonrelativistic quantum mechanics the commutator of the space coordinate q and the momentum p is non-zero:

$$\hbar \to 1,$$

$$[q, p] = i.$$

The quantization of the scalar field is similar. We replace the coordinate q by the field, the momentum p by the momentum of the field. The commutator is proportional to the delta function:

$$q \to \varphi(x, t),$$

$$p \to \pi(y, t),$$

$$[\varphi(\vec{x},t),\dot{\varphi}(\vec{y},t)] = [\varphi(\vec{x},t),\pi(\vec{y},t)] = i\delta^3(\vec{x}-\vec{y}),$$
$$[\varphi(\vec{x},t),\varphi(\vec{y},t)] = 0,$$
$$[\dot{\varphi}(\vec{x},t),\dot{\varphi}(\vec{y},t)] = [\pi(\vec{x},t),\pi(\vec{y},t)] = 0.$$

Using these commutation relations, one can calculate the commutator of the four-momentum and of the field:

$$i[P^\mu,\varphi(\vec{x},t)] = \partial^\mu\varphi(\vec{x},t).$$

Similar to the harmonic oscillator the field can be described by annihilation and creation operators:

$$\varphi(x,t) = \int \frac{d^3p}{\sqrt{(2\pi)^3 2\omega_p}}(a(p)\cdot e^{-ipx} + a^\dagger(p)\cdot e^{ipx}),$$

$$\omega_p = \sqrt{\vec{p}^2+m^2},$$

$$[a(p),a^\dagger(p')] = \delta^3(\vec{p}-\vec{p}'),$$
$$[a(p),a(p')] = [a^\dagger(p),a^\dagger(p')] = 0.$$

The commutation relations are invariant under Lorentz transformations:

$$\frac{d^3p}{2\omega_p} = d^4p\cdot\delta(p^2-m^2)\cdot\theta(p^0).$$

The Hamilton operator and the momentum operator can be written as functions of the creation and annihilation operators:

$$H = \int d^3p\cdot\omega_p\cdot\left(a^\dagger\cdot a + \frac{1}{2}\right),$$

$$[H,a] = -\omega_p\cdot a, \quad [H,a^\dagger] = +\omega_p\cdot a^\dagger,$$

$$\vec{P} = \int d^3p\cdot\vec{p}\cdot\left(a^\dagger\cdot a + \frac{1}{2}\right).$$

If an annihilation operator is applied to an energy eigenstate, the energy is reduced. A creation operator increases the energy:

$$Ha(p)|E\rangle = (E - \omega_p) \cdot a(p)|E\rangle,$$
$$Ha^\dagger(p)|E\rangle = (E + \omega_p) \cdot a^\dagger(p)|E\rangle.$$

All states of an oscillator can be generated by applying the creation operator on the ground state. In the field theory the ground state is called the *vacuum*. An annihilation operator annihilates the vacuum:

$$a(\vec{p})|0\rangle = 0.$$

If a creation operator is applied on the vacuum, a particle is created:

$$a^\dagger(\vec{p})|0\rangle = |\vec{p}\rangle.$$

In a similar way one obtains a two-particle state:

$$|p_1, p_2\rangle = a^\dagger(p_1) \cdot a^\dagger(p_2)|0\rangle.$$

This state is symmetric with respect to the interchange of the two particles, thus they are bosons and obey the Bose statistics.

The vacuum matrix element of the Hamilton operator is infinite:

$$\langle 0|H|0\rangle = \langle 0| \int d^3p \cdot \omega_p \left(a^\dagger \cdot a + \frac{1}{2}\right)|0\rangle = \frac{1}{2}\int d^3p \cdot \omega_p = \infty.$$

This problem can be avoided if the definition of the product of operators is changed — the creation operators must always be left of the annihilation operators (*normal ordering*):

$$aa^\dagger \Rightarrow a^\dagger a,$$
$$H \Rightarrow \int d^3\vec{p} \cdot \omega_p \cdot (a^\dagger a),$$
$$\vec{P} \Rightarrow \int d^3\vec{p} \cdot \vec{p} \cdot (a^\dagger a).$$

In coordinate space the normal ordering is described by a colon:

$$H(x) = \frac{1}{2}(: \pi^2 : + : (\vec{\nabla}\varphi)^2 : + m^2 : \varphi^2 :).$$

A field can be decomposed into a positive and a negative frequency part. The positive frequency part contains only annihilation operators, the negative frequency part only creation operators:

$$\varphi^{+}(x) = \int \frac{d^3\vec{p}}{\sqrt{(2\pi)^3 2\omega_p}} a(\vec{p}) \cdot e^{-ipx},$$

$$\varphi^{-}(x) = \int \frac{d^3\vec{p}}{\sqrt{(2\pi)^3 2\omega_p}} a^{\dagger}(\vec{p}) \cdot e^{ipx}\varphi.$$

The positive frequency part applied to the vacuum gives zero:

$$\varphi^{+}(x)|0\rangle = 0.$$

Complex Scalar Fields

A complex scalar field is a function of two independent real scalar fields:

$$\phi = \frac{1}{\sqrt{2}}(\varphi_1 + i\varphi_2).$$

Complex scalar fields can be used to describe charged scalar particles, e.g. the charged π mesons. Here are the Lagrange density, the momentum and the Hamilton density of a complex scalar field:

$$L = :\partial_\mu \phi^* \partial^\mu \phi: - m^2 :\phi^* \phi:,$$

$$\pi = \partial_0 \phi^*,$$

$$H(x) = :\pi^* \pi: + :\vec{\nabla}\phi^* \nabla\phi: + m^2 :\phi^* \phi: .$$

Similar to the real scalar case the commutation relation of the field and its momentum is given by the delta function:

$$[\phi(\vec{x}, t), \pi(\vec{y}, t)] = i\delta^3(\vec{x} - \vec{y}).$$

For complex scalar fields the Lagrange density remains invariant under a phase transformation:

$$\phi \Rightarrow \exp(-i\lambda) \cdot \phi.$$

Such a symmetry transformation leads to the conserved charge Q:

$$Q = i \int d^3x(:\phi^* \cdot \dot{\phi}: - :\dot{\phi}^* \cdot \phi:),$$

$$[Q, \phi] = -\phi,$$

$$[Q, \phi^*] = +\phi^*.$$

The charge Q is the space integral of the time component of the current:

$$j_\mu = i(:\phi^* \cdot \partial_\mu \phi: - :\partial_\mu \phi^* \cdot \phi:).$$

The complex field is a function of two different annihilation and creation operators:

$$\phi(x) = \int \frac{d^3p}{\sqrt{(2\pi)^3 2\omega_p}} (a(p) \cdot e^{-ipx} + b^\dagger(p) \cdot e^{ipx}),$$

$$\omega_p = \sqrt{\vec{p}^2 + m^2}.$$

Here are the commutation relations, the Hamilton operator H and the charge operator Q:

$$[a(\vec{p}), a^\dagger(\vec{p}')] = [b(\vec{p}), b^\dagger(\vec{p}')] = \delta^3(\vec{p} - \vec{p}'),$$

$$H = \int d^3\vec{p} \cdot \omega_p \cdot (a^\dagger a + b^\dagger b),$$

$$Q = \int d^3\vec{p} \cdot (a^\dagger a - b^\dagger b).$$

The Lagrange density of a complex field is also invariant under a new symmetry, the charge conjugation C. This symmetry interchanges particles and antiparticles:

$$C\phi C^{-1} = \phi^*,$$

$$CQC^{-1} = -Q.$$

Covariant Commutation Relations

We have discussed the commutation relations of the scalar fields, defined for the same time. For free fields one can also determine the commutation relations for different times. We consider a real field and decompose it into the positive and negative frequency parts:

$$\varphi = \varphi^+ + \varphi^-.$$

The commutator for arbitrary times can be calculated with the help of the creation and annihilation operators:

$$[\varphi^+(x), \varphi^-(y)] = \iint \frac{d^3q \cdot d^3p}{(2\pi)^3 2\sqrt{\omega_p \omega_q}} [a(\vec{q}), a^\dagger(\vec{p})] \cdot e^{-i(qx-py)}$$

$$= \frac{1}{(2\pi)^3} \int \frac{d^3\vec{q}}{2\omega_q} \cdot e^{-iq(x-y)} = i\Delta^+(x-y),$$

$$[\varphi^-(x), \varphi^+(y)] = i\Delta^-(x-y).$$

Both functions are solutions of the Klein–Gordon equation:

$$(\partial_\mu \partial^\mu + m^2)\Delta^{+,-}(x) = 0.$$

The commutator of two fields is the sum of the two functions:

$$[\varphi(x), \varphi(y)] = i\Delta^+(x-y) + i\Delta^-(x-y) = i\Delta(x-y).$$

This function had been calculated for the first time by P. Jordan and W. Pauli — it is called the *Pauli–Jordan function.* It vanishes for space-like distances and can be written as an integral:

$$\Delta(x) = \frac{-1}{(2\pi)^3} \int \frac{d^3\vec{p}}{\omega_p} \cdot \sin(px).$$

This integral can be written in a different form:

$$\frac{d^3p}{2\omega_p} = d^4p \cdot \theta(p_0) \cdot \delta(p^2 - m^2),$$

$$\Delta(x) = -\frac{i}{(2\pi)^3} \int d^4p \cdot e^{-ipx} \delta(p^2 - m^2) \cdot \varepsilon(p),$$

$$\varepsilon(x) = \Theta(x) - \Theta(-x),$$
$$\Theta(x) = 1 \quad (x > 0),$$
$$\Theta(x) = 0 \quad (x < 0).$$

The time derivative of the Pauli–Jordan function at time 0 is a delta function:

$$\partial^0 \Delta(x)|_{x_0=0} = -\delta^3(\vec{x}).$$

In coordinate space the Pauli-Jordan function is given by a Bessel function, which in the vicinity of the light cone can be written as follows:

$$\Delta(x) \approx -\frac{1}{2\pi}\varepsilon(x^0) \cdot \delta(x^2) + \frac{m^2}{8\pi}\varepsilon(x^0) \cdot \theta(x^2).$$

These functions can be written as a line integral in the complex momentum space:

$$\Delta^{\pm}(x) = -\frac{1}{(2\pi)^4} \int_{C\pm} \frac{d^4p \cdot e^{-ipx}}{p^2 - m^2},$$
$$\Delta(x) = -\frac{1}{(2\pi)^4} \int_{C} \frac{d^4p \cdot e^{-ipx}}{p^2 - m^2}.$$

The integration paths are given in Fig. 6.1.

Now we define the time-ordered product of two field operators as a product where the sequence of the factors depends on time — the factor with earlier time is left of the factor with later time:

$$T\varphi(x) \cdot \varphi(y) = \theta(x^0 - y^0) \cdot \varphi(x) \cdot \varphi(y) + \theta(y^0 - x^0) \cdot \varphi(y) \cdot \varphi(x).$$

The vacuum expectation value of this product is the Feynman propagator:

$$i\Delta_F(x - y) = \langle 0|T\varphi(x) \cdot \varphi(y)|0\rangle,$$
$$\Delta_F(x) = \theta(x^0) \cdot \Delta^+(x) - \theta(-x_0) \cdot \Delta^-(x).$$

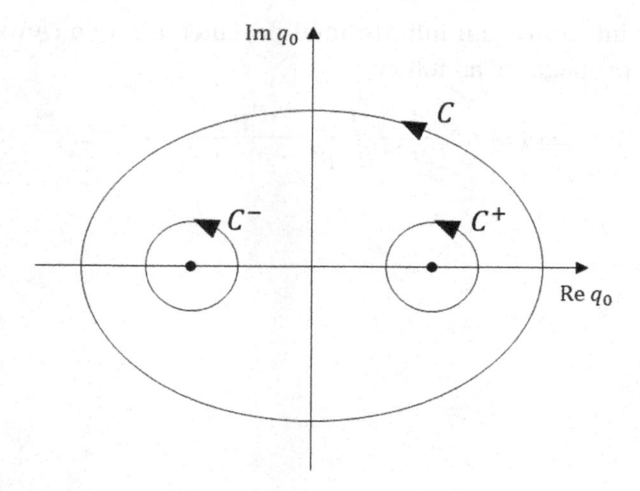

Fig. 6.1. The integration paths in complex momentum space.

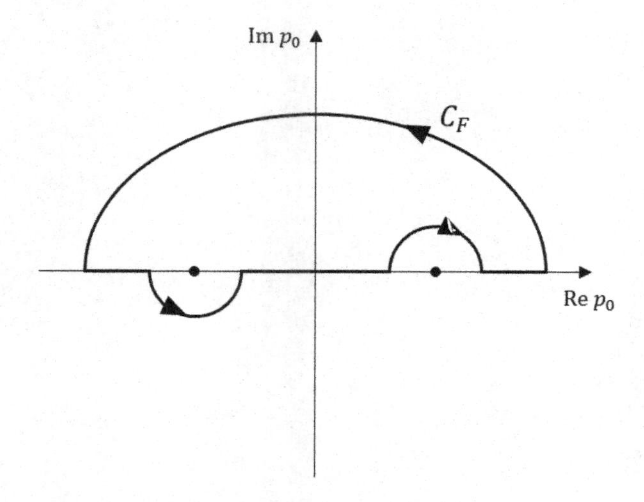

Fig. 6.2. Path of integration of the Feynman propagator.

Here is the Feynman propagator as an integral in momentum space:

$$\Delta_F(x) = \frac{1}{(2\pi)^4} \int_{C_F} \frac{d^4p}{p^2 - m^2} e^{-ipx}.$$

The integration path is described in Fig. 6.2.

If one introduces an infinitesimal parameter ε, one can write the Feynman propagator as follows:

$$\Delta_F(x) = \frac{1}{(2\pi)^4} \int \frac{d^4p}{p^2 - m^2 + i\varepsilon} \cdot e^{-ipx}.$$

Chapter 7

Free Spinor Fields

Electrons have an internal angular momentum, the *spin*, which is a quantum phenomenon. There is no classical analogue. The wave function of the electron consists of two functions:

$$\Psi = \begin{pmatrix} \psi_1 \\ \psi_2 \end{pmatrix}.$$

The two wave functions describe the electron with the spin (+1/2) or (–1/2). This wave function is a nonrelativistic spinor.

Dirac Equation

The Schrödinger equation is asymmetric with respect to space and time. In a relativistic wave equation, space and time must appear in a symmetric way. This is true for the Klein–Gordon equation:

$$\left(\frac{\partial^2}{\partial t^2} - \vec{\nabla}^2 + m^2 \right) \phi(\vec{x}, t) = 0.$$

It does not change under a Lorentz transformation. But it is not possible to construct a positive probability density as a function of the wave function. Thus the Klein–Gordon equation is not a relativistic generalization of the Schrödinger equation.

Paul Dirac studied in 1928 another equation, in which space and time also appear symmetrically:

$$i\hbar\frac{\partial\psi}{\partial t} = H\psi = \frac{\hbar}{i}\left(\alpha_1\frac{\partial\psi}{\partial x^1} + \alpha_2\frac{\partial\psi}{\partial x^2} + \alpha_3\frac{\partial\psi}{\partial x^3}\right) + m\cdot\beta\cdot\psi.$$

The coefficients in this equation cannot be numbers, since in this case the equation would not be invariant under rotations. Dirac assumed that the coefficients are matrices.

The wave function has several components — the number of components is determined by the matrices. For example, if the matrices are (4×4)-matrices, the wave function has 4 components:

$$\psi = \begin{pmatrix}\psi_1\\ \psi_2\\ \psi_3\\ \psi_4\end{pmatrix}, \quad \psi^\dagger = (\psi_1^*, \psi_2^*, \psi_3^*, \psi_4^*).$$

To figure out the number of dimensions necessary for the wave function, we first note that the wave function must obey the Klein–Gordon equation. Therefore the matrices must satisfy certain algebraic relations, which depend on the so-called anti-commutator of two operators:

$$\{A, B\} = A\cdot B + B\cdot A,$$
$$\{\alpha_i, \alpha_k\} = 2\delta_{ik},$$
$$\{\alpha_i, \beta\} = 0.$$

These matrices are $(N\times N)$ Hermitian matrices. The eigenvalues of the matrices must be $(+1)$ or (-1) and the matrices must be traceless. This implies, that N must be an even number. The possibility of $N = 2$ is excluded, since there are only three anti-commuting matrices, the three Pauli matrices:

$$\sigma_1 = \begin{pmatrix}0 & 1\\ 1 & 0\end{pmatrix}, \quad \sigma_2 = \begin{pmatrix}0 & -i\\ i & 0\end{pmatrix}, \quad \sigma_3 = \begin{pmatrix}1 & 0\\ 0 & -1\end{pmatrix}.$$

Thus the smallest dimension is $N = 4$. Here is one possibility for the (4×4)-matrices, using the Pauli matrices and the identity matrix as components:

$$\alpha_i = \begin{bmatrix} 0 & \sigma_i \\ \sigma_i & 0 \end{bmatrix}, \quad \beta = \begin{bmatrix} I & 0 \\ 0 & -I \end{bmatrix}.$$

The probability density and the associated current are bilinear in the wave function:

$$\rho = \psi^\dagger \psi,$$

$$j^k = \psi^\dagger \alpha_k \psi, k = 1, 2, 3.$$

These four quantities form a four-vector, the current, which is conserved:

$$\frac{\partial}{\partial t}\rho + \vec{\nabla}\vec{j} = 0.$$

The first two components of the four-component wave function describe the spin of the particle, the other two components the spin of the corresponding antiparticle. If the particle is the electron, the antiparticle is the positron:

$$\psi = \begin{pmatrix} e^-(+1/2) \\ e^-(-1/2) \\ e^+(+1/2) \\ e^+(-1/2) \end{pmatrix}.$$

Often one uses other matrices, defined by:

$$\gamma^0 = \beta,$$

$$\gamma^i = \beta\alpha_i,$$

$$\{\gamma^\mu, \gamma^\nu\} = g^{\mu\nu}.$$

Using those matrices, the Dirac equation takes a different form:

$$i\gamma^\mu \partial_\mu \psi(x) - m\psi(x) = 0,$$

$$(\gamma^\mu p_\mu - m) \cdot \psi(p) = 0.$$

The corresponding Lagrange density is:

$$L = \bar{\psi}(i\gamma^\mu \partial_\mu - m)\psi.$$

The Dirac wave function obeys also the Klein–Gordon equation:

$$\partial^\mu \partial_\mu \psi + m^2 \psi = 0.$$

This is obtained, if one uses the Dirac representation of the matrices, in which the three Pauli matrices appear:

$$\beta = \gamma_0 = \begin{pmatrix} I & 0 \\ 0 & -I \end{pmatrix}, \quad \vec{\gamma} = \begin{pmatrix} 0 & \vec{\sigma} \\ -\vec{\sigma} & 0 \end{pmatrix}.$$

We define another matrix:

$$\gamma^5 = i\gamma^0\gamma^1\gamma^2\gamma^3.$$

This matrix has the following properties:

$$(\gamma^5)^\dagger = \gamma^5,$$
$$(\gamma^5)^2 = I,$$
$$\{\gamma^5, \gamma^\mu\} = 0.$$

Here is the matrix in the Dirac representation:

$$\gamma^5 = \begin{pmatrix} -1 & 0 & 0 & 0 \\ 0 & -1 & 0 & 0 \\ 0 & 0 & 1 & 0 \\ 0 & 0 & 0 & 1 \end{pmatrix}.$$

Often it is necessary to calculate traces of products of gamma matrices. We mention some useful relations:

$$\mathrm{tr}(\gamma^\mu\gamma^\nu) = 4g^{\mu\nu},$$
$$\mathrm{tr}(\gamma^\mu\gamma^\nu\gamma^\rho\gamma^\sigma) = 4(g^{\mu\nu}g^{\rho\sigma} - g^{\mu\rho}g^{\nu\sigma} + g^{\mu\sigma}g^{\nu\rho}),$$
$$\mathrm{tr}(\gamma^5) = 0,$$

$$\mathrm{tr}(\gamma^\mu\gamma^\nu\gamma^5) = 0,$$

$$\mathrm{tr}(\gamma^\mu\gamma^\nu\gamma^\rho\gamma^\sigma\gamma^5) = -4i\varepsilon^{\mu o\rho\sigma},$$

$$\gamma^\mu\gamma^\nu\gamma_\mu = -2\gamma_\nu,$$

$$\gamma^\mu\gamma^\nu\gamma^\rho\gamma_\mu = 4g^{\nu\rho},$$

$$\gamma^\mu\gamma^\nu\gamma^\rho\gamma^\sigma\gamma_\mu = -2\gamma^\sigma\gamma^\rho\gamma^\nu.$$

The adjoint Dirac spinor is defined as follows:

$$\bar{\psi} = \psi^\dagger\gamma^0.$$

Here is the associated Dirac equation:

$$i\partial_\mu\bar{\psi}(x)\cdot\gamma^\mu + m\cdot\bar{\psi}(x) = 0.$$

The Dirac wave function is not an observable, but the following bilinear quantities are:

$$\bar{\psi}\psi, \quad \bar{\psi}\gamma_5\psi, \quad \bar{\psi}\gamma_\mu\psi, \quad \bar{\psi}\gamma_\mu\gamma_5\psi, \quad \bar{\psi}\sigma_{\mu\nu}\psi.$$

These are a scalar, a pseudo-scalar, the four components of a vector, the four components of a pseudo-vector and the six components of an asymmetric tensor.

Richard Feynman introduced the "slash" notation. A four-vector can be written as a matrix:

$$\gamma^\mu\cdot a_\mu = \not{a},$$

$$\gamma^\mu\cdot p_\mu = \not{p} = \begin{pmatrix} E & -\vec{\sigma}\cdot\vec{p} \\ \vec{\sigma}\cdot\vec{p} & -E \end{pmatrix}.$$

Here is the Dirac equation in the slash notation:

$$(i\not{\partial} - m)\cdot\psi = 0,$$

$$(\not{p} - m)\cdot\psi = 0.$$

The vector current, which is bilinear in the Dirac field, is conserved:

$$\partial^\mu\bar{\psi}\gamma_\mu\psi = 0.$$

The charge Q is given by the space integral:

$$Q = \int d^3x \psi^\dagger \psi.$$

The axial vector current is only conserved, if the fermion mass is zero:

$$\partial^\mu \bar{\psi} \gamma_\mu \gamma_5 \psi = 2im \bar{\psi} \gamma_5 \psi.$$

The Lorentz transformation of a spinor is described by a matrix S, which is a function of the Lorentz matrix:

$$x'^\mu = \Lambda^\mu_\nu \cdot x^\nu,$$

$$\psi' = S(\Lambda) \cdot \psi.$$

The matrix S can be calculated from the relation:

$$S^{-1} \gamma^\mu S = \Lambda^{\mu\nu} \gamma_\nu.$$

We consider two special cases.

1. Infinitesimal rotation around the z axis:

$$\psi' \cong \left(1 + i\frac{\delta}{2} \begin{pmatrix} \sigma_3 & 0 \\ 0 & \sigma_3 \end{pmatrix}\right) \psi.$$

2. Rotation of the time coordinate and the z coordinate:

$$\begin{pmatrix} t' \\ z' \end{pmatrix} = \begin{pmatrix} \cosh\xi & \sinh\xi \\ -\sinh\xi & \cosh\xi \end{pmatrix} \begin{pmatrix} t \\ z \end{pmatrix},$$

$$\psi' \cong \left(1 + \frac{\xi}{2} \begin{pmatrix} 0 & \sigma_3 \\ \sigma_3 & 0 \end{pmatrix}\right) \psi.$$

The transformation of the adjoint spinor is also given by the matrix S:

$$\psi \to S \cdot \psi,$$

$$\bar{\psi} \to \bar{\psi} \cdot S^{-1}.$$

Quantization

The momentum of a Dirac field is given by

$$\pi = \frac{\partial L}{\partial \dot{\psi}} = i\bar{\psi}\gamma^0 = i\psi^\dagger.$$

The Hamilton operator is determined by the space integral of the Hamilton density:

$$H = \int d^3x \cdot (\psi^\dagger i\partial_0\psi).$$

Recall the commutation relations of a scalar field:

$$[\pi(\vec{x}, t), \phi(\vec{x}\,', t)] = -i\delta^3(\vec{x} - \vec{x}\,').$$

For a spinor field the commutation relations are replaced by anti-commutation relations:

$$\{\psi_\alpha(\vec{x}, t), \pi_\beta(\vec{x}\,', t)\} = i\delta(\vec{x} - \vec{x}\,')\delta_{\alpha\beta},$$
$$\{\psi_\alpha(\vec{x}, t), \psi^\dagger_\beta(\vec{x}\,', t)\} = \delta(\vec{x} - \vec{x}\,')\delta_{\alpha\beta},$$
$$\{\psi_\alpha(\vec{x}, t), \psi_\beta(\vec{x}\,', t)\} = 0,$$
$$\{\psi^\dagger_\alpha(\vec{x}, t), \psi^\dagger_\beta(\vec{x}\,', t)\} = 0.$$

The spinor field can also be written as a function of creation and annihilation operators:

$$\psi(x) = \sum_{r=1}^{2} \int d^3p \sqrt{\frac{m}{(2\pi)^3 p^0}} (c_r \cdot u_r \cdot e^{-ipx} + d_r^\dagger \cdot v_r \cdot e^{ipx}),$$

$$\bar{\psi}(x) = \sum_{r=1}^{2} \int d^3p \sqrt{\frac{m}{(2\pi)^3 p^0}} (d_r \cdot \bar{v}_r \cdot e^{-ipx} + c_r^\dagger \cdot \bar{u}_r \cdot e^{ipx}),$$

$$\{c_r(\vec{p}), c_s^\dagger(\vec{k})\} = \{d_r(\vec{p}), d_s^\dagger(\vec{k})\} = \delta_{rs}\delta^3(\vec{p} - \vec{k}).$$

Here are the Hamilton operator and the charge operator:

$$H = \sum_r \int d^3\vec{p} \cdot E_p(c_r^\dagger c_r + d_r^\dagger d_r),$$

$$Q = -e \cdot \sum_r \int d^3\vec{p}(c_r^\dagger c_r - d_r^\dagger d_r).$$

We consider a solution of the Dirac equation in the nonrelativistic limit and with vanishing momentum. Here are the four components of the spinor:

$$u_1(0) = \begin{pmatrix} 1 \\ 0 \\ 0 \\ 0 \end{pmatrix}, \quad u_2(0) = \begin{pmatrix} 0 \\ 1 \\ 0 \\ 0 \end{pmatrix}, \quad v_1(0) = \begin{pmatrix} 0 \\ 0 \\ 1 \\ 0 \end{pmatrix}, \quad v_2(0) = \begin{pmatrix} 0 \\ 0 \\ 0 \\ 1 \end{pmatrix}.$$

The energy of the first two components is positive — they describe the fermion with the two spins, e.g. the electron. The energy of the other components is negative, thus these components describe the anti-fermion, e.g. the positron.

Using the anti-commutation relations for the creation and annihilation operators, one can calculate the anti-commutator of the free spinor field at arbitrary times:

$$\{\psi_\alpha(x), \bar{\psi}_\beta(y)\} = iS(x-y) = i(i\partial\!\!\!/ + m)_{\alpha\beta}\Delta(x-y),$$

$$S(x) = \frac{-1}{(2\pi)^4} \int dp \cdot e^{-ipx} \frac{p\!\!\!/ + m}{p^2 - m^2}.$$

The time-ordered product for spinor fields is defined as follows:

$$T(\psi(x) \cdot \bar{\psi}(y)) = \theta(x^0 - y^0) \cdot \psi(x) \cdot \bar{\psi}(y) - \theta(y^0 - x^0) \cdot \bar{\psi}(y) \cdot \psi(x).$$

The negative sign of the second term follows from the anti-commutation relations.

The Feynman propagator is determined by the time-ordered product of two spinor fields:

$$iS_F(x-y) = \langle 0|\, T\psi(x)\cdot\bar{\psi}(y)\,|0\rangle\,,$$

$$S_F(x) = \frac{1}{(2\pi)^4}\int dp\cdot e^{-ipx}\frac{\not{p}+m}{p^2-m^2+i\varepsilon}.$$

The Dirac operator, applied to the Feynman propagator, gives the four-dimensional delta function:

$$(i\not{\partial}-m)S_F(x) = \delta^4(x).$$

Chapter 8

Free Vector Fields

Vector fields are important for the description of the fundamental interactions. These are generated by the exchange of vector particles. The electromagnetic interaction between charged particles is due to the exchange of photons, which are vector particles. The strong interactions between the quarks inside a proton are generated by the exchange of vector gluons. The weak interaction is due to the exchange of weak vector bosons.

The photon, the eight gluons and the three weak bosons are described by quantized vector fields. The photon and the gluons are massless, but the two charged W bosons and the neutral Z boson have large masses:

$$M(\mathrm{W}) \sim 80\,\mathrm{GeV},$$
$$M(\mathrm{Z}) \sim 91\,\mathrm{GeV}.$$

A vector field A transforms as a vector under Lorentz transformations:

$$x'^{\mu} = \Lambda^{\mu}_{\nu} \cdot x^{\nu},$$
$$A'^{\mu} = \Lambda^{\mu}_{\nu} \cdot A^{\nu}.$$

The electromagnetic vector field cannot be observed, only the associated field strengths. Those are given by the antisymmetric field strength tensor F, which contains the three components of the electric field strength E and the three components of the magnetic field strength H:

$$F^{\mu\nu} = \partial^\mu A^\nu - \partial^\nu A^\mu \Rightarrow \begin{pmatrix} 0 & -E_1 & -E_2 & -E_3 \\ E_1 & 0 & -H_3 & H_2 \\ E_2 & H_3 & 0 & -H_1 \\ E_3 & -H_2 & H_1 & 0 \end{pmatrix}.$$

The vector field is not uniquely determined by the field strength tensor. If a vector field is changed by the addition of the gradient of a scalar function, the field strength tensor remains invariant. Such a change is called a *gauge transformation*:

$$A'^\mu = A^\mu + \partial^\mu \alpha,$$
$$F'^{\mu\nu} = F^{\mu\nu}.$$

By a gauge transformation one can always arrange that the divergence of the vector field vanishes (Lorenz gauge condition):

$$\partial_\mu A^\mu = 0.$$

The Lagrange density of a free electromagnetic field is given by the square of the field strength tensor:

$$L = -\frac{1}{4} F_{\mu\nu} F^{\mu\nu}.$$

Thus one obtains the field equation:

$$\partial_\mu F^{\mu\nu} = 0.$$

By a suitable gauge transformation one can always arrange that the time component of the vector field and the divergence of the three space components vanish — Coulomb gauge:

$$A^0 = 0,$$
$$\vec{\nabla}\vec{A} = 0.$$

So we can see that the space components of the vector field obey the Klein–Gordon equation:

$$\partial_\mu \partial^\mu \vec{A} = 0.$$

The derivatives of the vector field define the electric and magnetic field strength:

$$\vec{H} = \vec{\nabla} \times \vec{A},$$

$$\vec{E} = -\frac{\partial \vec{A}}{\partial t}.$$

We consider the example of a plane wave, given by the momentum p:

$$A^{\mu} \sim e^{-ipx} \cdot \varepsilon^{\mu}(p).$$

This four-vector is the polarization vector. The Lorenz gauge condition implies:

$$p_{\mu}\varepsilon^{\mu}(p) = 0.$$

In the Coulomb gauge the time component of the polarization vector vanishes:

$$\vec{\varepsilon} \cdot \vec{p} = 0.$$

If the momentum is parallel to the z-axis, the two polarization vectors are given by two four vectors:

$$\varepsilon_1 = \begin{pmatrix} 0 \\ 1 \\ 0 \\ 0 \end{pmatrix}, \quad \varepsilon_2 = \begin{pmatrix} 0 \\ 0 \\ 1 \\ 0 \end{pmatrix}.$$

The Lagrange density given above does not allow a canonical quantization. For this reason one uses often the following Lagrange density:

$$L = -\frac{1}{2}(\partial_{\nu} A_{\mu}) \cdot (\partial^{\nu} A^{\mu}).$$

In this case the momentum of the field is given by the time derivative of the field:

$$\pi^{\mu} = \frac{\partial L}{\partial(\partial_0 A_{\mu})} = -\partial^0 A^{\mu}.$$

Here are the equal-time commutation relations of the field and the momentum:

$$[A_\mu(\vec{x},t), \pi_\nu(\vec{y},t)] = -ig_{\mu\nu}\delta^3(\vec{x}-\vec{y}),$$
$$[A_\mu(\vec{x},t), A_\nu(\vec{y},t)] = [\pi_\mu(\vec{x},t), \pi_\nu(\vec{y},t)] = 0.$$

One can also calculate the covariant commutation relations, which are similar to those for a scalar field:

$$[A_\mu(x), A_\nu(y)] = -ig_{\mu\nu}\Delta(x-y).$$

The Feynman propagator is the vacuum expectation value of the time-ordered product of two field operators:

$$\langle 0|T(A_\mu(x)A_\nu(y))|0\rangle = -ig_{\mu\nu}\cdot\Delta_F(x-y).$$

The field operator can be written as a function of creation and annihilation operators:

$$A_\mu = \int \frac{d^3p}{2p_0(2\pi)^3}\sum_\lambda (a_\lambda\cdot\varepsilon_\mu^{(\lambda)}\cdot e^{ipx} + a_\lambda^\dagger\cdot\varepsilon_\mu^{(\lambda)}\cdot e^{-ipx}) = A_\mu^+ + A_\mu^-.$$

The creation and annihilation operators obey these commutation relations:

$$[a_\lambda(p), a_{\lambda'}^\dagger(p')] = \delta_{\lambda\lambda'}2p^0(2\pi)^3\delta^3(\vec{p}-\vec{p}').$$

A massless vector particle can either be polarized in the direction of the momentum (right-handed polarization) or in the opposite direction (left-handed polarization).

The Lagrange density of a vector field, which describes a massless particle, e.g. the photon, does not have a mass term. However a mass term is needed for the description of a massive vector particle:

$$L = -\frac{1}{4}F_{\mu\nu}F^{\mu\nu} - \frac{M^2}{2}A_\mu A^\mu.$$

The corresponding field equation is called the *Proca equation*:

$$(\partial_\mu\partial^\mu + M^2)A^\nu = 0.$$

A massive vector particle has three polarizations: left-handed, right-handed and longitudinal polarization.

Chapter 9

Perturbation Theory

A scattering process is the transition from an initial state, e.g. an electron and a proton with fixed momenta, to a final state. If the final state is also an electron and a proton, it is an elastic scattering process. But if the energy of the electron is large enough, the proton can be excited to a delta resonance, which afterwards decays e.g. into a neutron and a charged pion. The final state is now a state with three particles — it is an inelastic scattering process.

Scattering processes are described by a unitary matrix, the scattering matrix S. It describes the transition from the initial state to the final state:

$$|f\rangle = S|i\rangle,$$
$$|\phi(t = +\infty)\rangle = S|\phi(t = -\infty)\rangle.$$

An exact calculation of the S-matrix is not possible, but it can be calculated approximately, if one uses the perturbation theory.

The Hamilton operator is a sum of the Hamilton operator of the free particles and of an operator describing the interaction:

$$H = H_0 + H_i.$$

The time evolution of a state in the interaction picture is given by the Hamilton operator of the interaction:

$$i\frac{\partial}{\partial t}|\psi(t)\rangle = H_i|\psi(t)\rangle,$$

$$|\psi(t)\rangle = U_i(t,t_0)\cdot|\psi(t_0)\rangle.$$

If we consider the time evolution of the unitary operator, its time derivative is given by the Hamilton operator of the interaction:

$$i\frac{\partial U_i(t,t_0)}{\partial t} = H_i(t)\cdot U(t,t_0),$$

$$U_i(t,t_0) = 1 - i\int_{t_0}^{t} dt' H_i(t')\cdot U_i(t',t_0).$$

Now we decompose the time into n small time differences:

$$t_0 < t_1 < t_2 < \ldots < t_n = t.$$

The unitary operator can be written as a product:

$$U_i(t,t_0) = U_i(t,t_{n-1})\cdot\ldots\cdot U_i(t_2,t_1)\cdot U_i(t_1,t_0),$$

$$U_i(t_{j+1},t_j) \approx 1 - i\int_{t_j}^{t_{j+1}} dt' H_i(t').$$

Thus we find a power series:

$$U(t,t_0) = 1 + (-i)\int_{t_0}^{t} dt_1 H_i(t_1) + (-i)^2\int_{t_0}^{t} dt_1\int_{t_0}^{t_1} dt_2 H_i(t_1)\cdot H_i(t_2)$$

$$+(-i)^3\int_{t_0}^{t} dt_1\int_{t_0}^{t_1} dt_2\int_{t_0}^{t_2} dt_3 H_i(t_1)\cdot H_i(t_2)\cdot H_i(t_3) + \cdots$$

$$= \sum_{n=0}^{\infty}(-i)^n\int_{t_0}^{t} dt_1\int_{t_0}^{t_1} dt_2\ldots\int_{t_0}^{t_n} dt_n H_i(t_1)\cdot\ldots\cdot H_i(t_n).$$

It is given by a sum of integrals:

$$I_n = \int_{t_0}^{t} dt_1\int_{t_0}^{t_1} dt_2\ldots\int_{t_0}^{t_{n-1}} dt_n H_i(t_1)\cdot H_i(t_2)\cdot\ldots\cdot H_i(t_n).$$

Such an integral can be rewritten using the theta function:

$$I_n = \int_{t_0}^{t} dt_1 \int_{t_0}^{t} dt_2 \ldots \int_{t_0}^{t} dt_n \theta(t_1 - t_2) \cdot \theta(t_2 - t_3) \cdot \ldots \cdot \theta(t_{n-1} - t_n)$$

$$\cdot H_i(t_1) \cdot H_i(t_2) \cdot \ldots \cdot H_i(t_n)$$

$$= \frac{1}{n!} \int_{t_0}^{t} dt_1 \int_{t_0}^{t} dt_2 \ldots \int_{t_0}^{t} dt_n T(H_i(t_1) \cdot H_i(t_2) \cdot \ldots \cdot H_i(t_n)).$$

Here the symbol T means the time ordering of the factors, which implies the summation over all permutations of the different times:

$$T(H_i(t_1) \cdot H_i(t_2) \cdot \ldots \cdot H_i(t_n))$$

$$= \sum_P \theta(t_{\alpha_1} - t_{\alpha_2}) \cdot \ldots \cdot \theta(t_{\alpha_{n-1}} - t_{\alpha_n}) \cdot H_i(t_{\alpha_1}) \cdot \ldots \cdot H_i(t_{\alpha_n}).$$

For the case $n = 2$ one obtains:

$$\int_{t_0}^{t} dt_1 \int_{t_0}^{t} dt_2 T(H_i(t_1) \cdot H_i(t_2)) = \int_{t_0}^{t} dt_1 \int_{t_0}^{t_1} dt_2 H_i(t_1) \cdot H_i(t_2)$$

$$+ \int_{t_0}^{t} dt_1 \int_{t_1}^{t} dt_2 H_i(t_2) \cdot H_i(t_1).$$

The operator U can now be written as follows:

$$U(t, t_0) = \sum_{n=0}^{\infty} \frac{1}{n!} (-i)^n \int_{t_0}^{t} dt_1 \int_{t_0}^{t} dt_2 \cdots \int_{t_0}^{t} dt_n T(H_i(t_1)$$

$$\cdot H_i(t_2) \cdot \ldots \cdot H_i(t_n))$$

$$= T \exp\left(-i \int_{t_0}^{t} H_i(t') \cdot dt' \right).$$

In the limits $t_0 \to -\infty$ and $t \to +\infty$ one obtains the S-matrix:

$$S = U(-\infty, +\infty) = \sum_{n=0}^{\infty} \frac{(-i)^n}{n!} \int_{-\infty}^{\infty} dt_1 \ldots \int_{-\infty}^{\infty} dt_n T(H_i(t_1)$$

$$\cdot H_i(t_2) \cdot \ldots \cdot H_i(t_n))$$

$$= T \exp\left(-i \int_{-\infty}^{+\infty} H_i(t') \cdot dt'\right).$$

The Hamilton operator for the interaction can be written as an integral of the Hamilton density:

$$H_i = \int \widetilde{H}_i(x) d^3x.$$

Thus we find for the S-matrix:

$$S = \sum_{n=0}^{\infty} \frac{1}{n!}(-i)^n \int_{-\infty}^{\infty} d^4x_1 \ldots \int_{-\infty}^{\infty} d^4x_n T(\widetilde{H}_i(x_1)$$

$$\cdot \widetilde{H}_i(x_2) \cdot \ldots \cdot \widetilde{H}_i(x_n))$$

$$= T \exp\left(-i \int_{-\infty}^{+\infty} \widetilde{H}_i(x) d^4x\right).$$

We obtain for a matrix element of the S-matrix in momentum space:

$$S \propto -i(2\pi)^4 \delta^4(p_f - p_i) M_{fi}.$$

Here p_i and p_f are the sums of the momenta of the incoming and outgoing particles. The delta function is necessary to conserve the energies and the momenta.

Matrix elements of the S-matrix were calculated by Richard Feynman using special diagrams, which were later called *Feynman diagrams*. The matrix element M can be calculated with such diagrams. The matrix element is an infinite sum:

$$M = \sum_{i=o}^{\infty} g^i M_i.$$

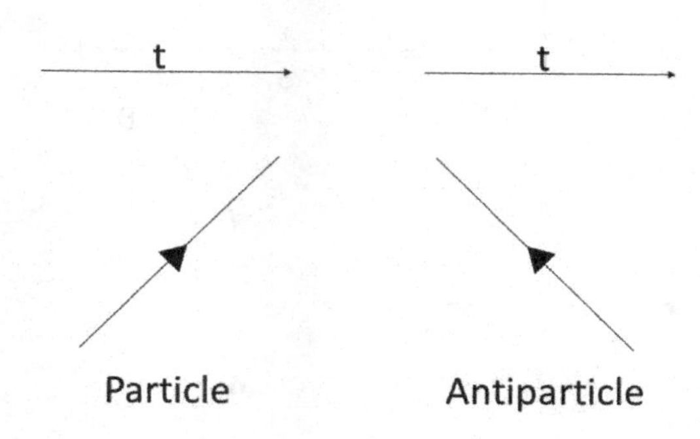

Fig. 9.1. Particles and antiparticles in space-time.

Every matrix element corresponds to a Feynman diagram. The parameter g is the coupling constant. For a small coupling constant, the infinite sum is approximately given by the first terms. In quantum electrodynamics the coupling constant e is determined by the fine structure constant:

$$\alpha = \frac{e^2}{4\pi} \approx 1/137 \approx 0.0073,$$

$$e \approx 0.303.$$

In a Feynman diagram an incoming particle is described by an arrow pointing in the direction of time (Fig. 9.1). An arrow, which describes an incoming antiparticle, points opposite to the direction of the time.

A simple diagram is the decay of a particle A into two particles B and C: A $\rightarrow$ B + C (Fig. 9.2).

The external lines in a Feynman diagram describe the incoming and outgoing particles. Vertices connect the external lines with the internal lines, which describe the virtual particles.

In Figs. 9.3 and 9.4, two types of scattering between two particles A and B are shown. The incoming particles produce a new virtual particle C, which decays into the particles A and B; or the new particle C is exchanged between A and B.

Important for the calculation of Feynman diagrams are the propagators of particles. Here are some examples.

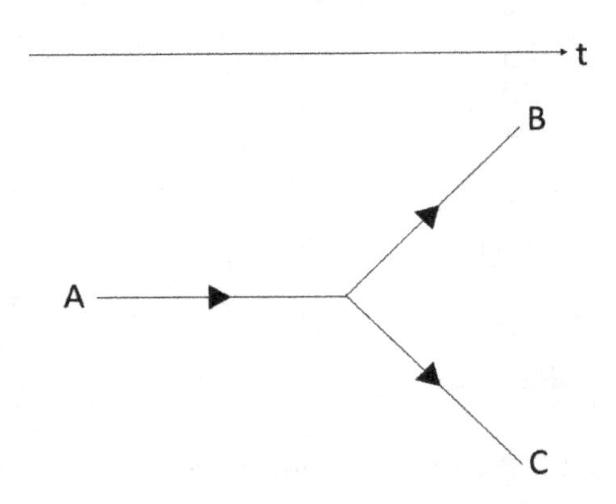

Fig. 9.2. The Feynman diagram for the decay A $\rightarrow$ B + C.

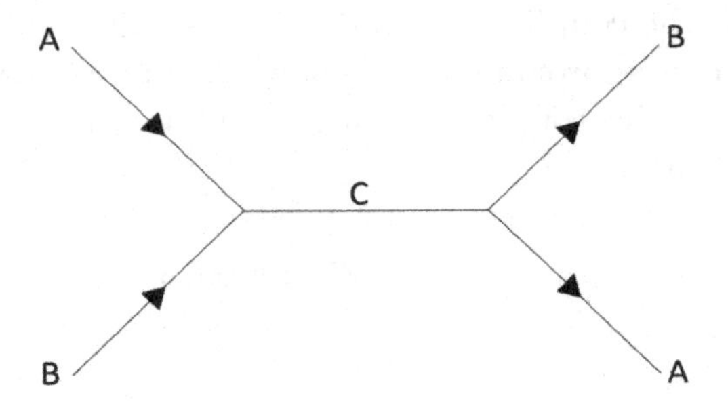

Fig. 9.3. The scattering of the particles A and B: a new virtual particle C is produced and decays immediately.

a) ***Propagator of a scalar particle***:

$$\frac{i}{q^2 - m^2}.$$

b) ***Propagator of a fermion***:

$$i\frac{\not{q} + m}{q^2 - m^2} = \frac{i}{\not{q} - m},$$

$$\not{q} = \gamma^\mu q_\mu.$$

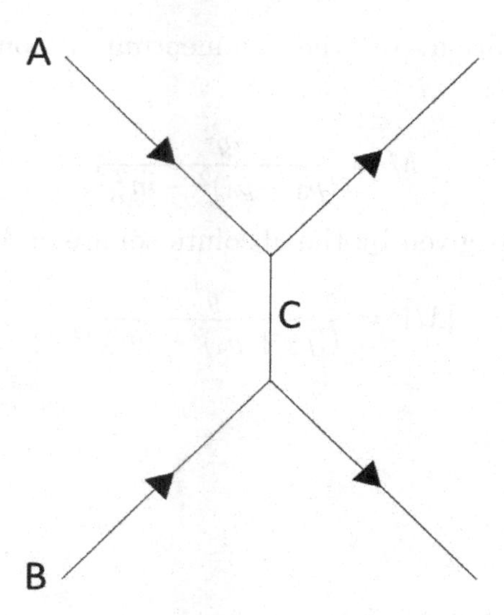

Fig. 9.4. The exchange of the new particle C describes the interaction between the particles A and B.

c) *Propagator of a photon:*

$$-\frac{i}{k^2}g^{\mu\nu}.$$

Example

A particle and its antiparticle annihilate and produce a scalar boson B, which decays into the particle and the antiparticle. The Feynman diagram is given by two vertices, which are proportional to the coupling constant g. The sum of the two momenta of the incoming particles must be equal to the momentum of the boson B — thus a delta function appears. The momentum of the particle B must be equal to the sum of the momenta of the outgoing particles — this gives a second delta function. Both delta functions are multiplied with the propagator of B:

$$g^2\frac{i}{q^2-m_B^2}(2\pi)^4\delta(p_1+p_2-q)\cdot(2\pi)^4\delta(q-p_3-p_4).$$

After integration of the momentum q one obtains the amplitude M:

$$M = \frac{ig^2}{(p_3 + p_4)^2 - m_B^2}.$$

The probability is given by the absolute square of M:

$$|M|^2 = \frac{g^4}{((p_3 + p_4)^2 - m_B^2)^2}.$$

Chapter 10
Quantum Electrodynamics

The theory of quantum electrodynamics (QED) combines electrodynamics, quantum mechanics and relativity theory. It describes the interaction of charged particles with the electromagnetic field.

The Lagrange density of a free complex scalar field describes charged scalar particles:

$$L_0(\phi, \partial\phi) = \partial_\mu \phi^* \partial^\mu \phi - m^2 \phi^* \phi.$$

It does not change under a phase transformation of the field:

$$\phi \to U\phi,$$
$$U = e^{ie\alpha}.$$

This phase transformation is a global gauge transformation. The generator of the transformation, the corresponding charge Q, is the space integral of the charge density, which is the time component of the current:

$$j_\mu = i \cdot (\phi^* \cdot \partial_\mu \phi - \partial_\mu \phi^* \cdot \phi),$$
$$\partial^\mu j_\mu = 0,$$
$$Q = \int j_0 d^3x.$$

A local gauge transformation is a transformation where the phase parameter depends on space and time:

$$\phi(x) \to U(x) \cdot \phi(x),$$
$$U(x) = e^{ie\alpha(x)}.$$

The derivative of the scalar field transforms under global gauge transformations in the same way as the field, but for a local gauge transformations there is an additional term:

$$\partial^\mu \phi(x) \to U(x) \cdot \partial^\mu \phi(x) + \phi(x) \cdot \partial^\mu U(x).$$

Thus the Lagrange density of a complex scalar field is not invariant under local gauge transformations. We have to replace the derivative of the scalar field by a covariant derivative, which transforms under local gauge transformations in the same way as the field. The covariant derivative contains a new vector field:

$$D^\mu \phi = (\partial^\mu + ieA^\mu)\phi,$$
$$D^\mu \phi(x) \to U(x) D^\mu \phi(x).$$

In quantum physics this vector field would describe a new vector particle. It can be identified with the photon, but in this case the Lagrange density must be expanded to the Lagrange density of scalar electrodynamics, given here with the field equations:

$$L = \frac{1}{2}(D_\mu \phi)^*(D_\mu \phi) - m^2 \phi^* \phi - \frac{1}{4} F_{\mu\nu} F^{\mu\nu},$$
$$F^{\mu\nu} = \partial^\mu A^\nu - \partial^\nu A^\mu,$$
$$\partial_\mu F^{\mu\nu} = j^\nu,$$
$$j^\mu = ie(\phi^* D^\mu \phi - (D^\mu \phi)^* \phi),$$
$$D_\mu D^\mu \phi + m^2 \phi = 0.$$

Thus the requirement of local gauge invariance introduces an interaction of the scalar field with the electromagnetic field.

There is no mass for the photon. One might introduce a photon mass by adding a mass term in the Lagrange density:

$$L = \frac{1}{2}(D_\mu\phi)^*(D_\mu\phi) - m^2\phi^*\phi - \frac{1}{4}F_{\mu\nu}F^{\mu\nu} - \frac{M^2}{2}A_\mu A^\mu.$$

But now the Lagrange density is not invariant under local gauge transformations — thus a photon mass is not allowed by local gauge invariance.

Now we consider the spinor electrodynamics. The Lagrange density of a free electron is the sum of the kinetic term and the mass term:

$$L = \bar{\psi}(i\gamma^\mu\partial_\mu - m_e)\psi.$$

This Lagrange density is invariant under global phase transformations:

$$\psi \to e^{ie\alpha}\psi,$$
$$\bar{\psi} \to e^{-ie\alpha}\bar{\psi}.$$

This symmetry is generated by the charge operator:

$$Q = \int d^3x \cdot j_o(x).$$

The charge Q is the space integral of the charge density, which is the time component of the conserved electric current:

$$j_\mu = \bar{\psi}\gamma_\mu\psi,$$
$$\partial^\mu j_\mu = 0.$$

Now we consider local gauge transformations:

$$\psi(x) \to e^{ie\alpha(x)}\psi(x).$$

The covariant derivative of the spinor field transforms under local gauge transformations in the same way as the spinor field:

$$\partial_\mu \rightarrow D_\mu = \partial_\mu + ieA_\mu,$$
$$\psi \rightarrow e^{ie\alpha(x)}\psi,$$
$$D_\mu\psi \rightarrow e^{ie\alpha(x)}D_\mu\psi.$$

We obtain the Lagrange density of quantum electrodynamics:

$$L = \bar{\psi}(i\gamma^\mu D_\mu - m)\psi - \frac{1}{4}F_{\mu\nu}F^{\mu\nu}$$
$$= \bar{\psi}(i\gamma^\mu \partial_\mu - m_e)\psi - \frac{1}{4}F_{\mu\nu}F^{\mu\nu} - e\bar{\psi}\gamma^\mu\psi \cdot A.$$

Here are the field equations:

$$(i\gamma^\mu\partial_\mu - m)\psi = e\cdot \not{A} \cdot \psi,$$
$$\frac{\partial F^{\mu\nu}}{\partial x^\nu} = e \cdot \bar{\psi}\gamma^\mu\psi.$$

The parameter e is the electromagnetic coupling constant, given by the fine-structure constant. This constant is a dimensionless number, which has been determined by many experiments:

$$\alpha = \frac{e^2}{4\pi} \cong \frac{1}{137.036},$$
$$e \approx 0.3028.$$

Quantum electrodynamics describes the interaction of electrons and photons. The forces between two electrons are generated by the exchange of virtual photons (Fig. 10.1).

The interaction of the electron with the photon is given by the product of the electromagnetic current and the photon field:

$$L_W \propto e \cdot \bar{\psi} \cdot \not{A} \cdot \psi.$$

The corresponding Feynman diagram is a vertex (Fig. 10.2).

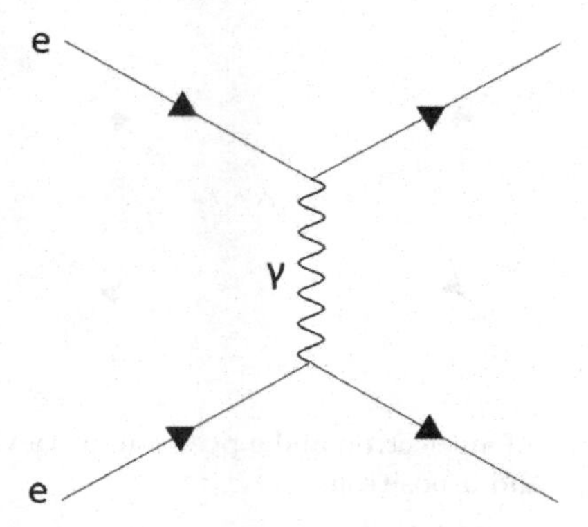

Fig. 10.1. The exchange of a virtual photon between two electrons.

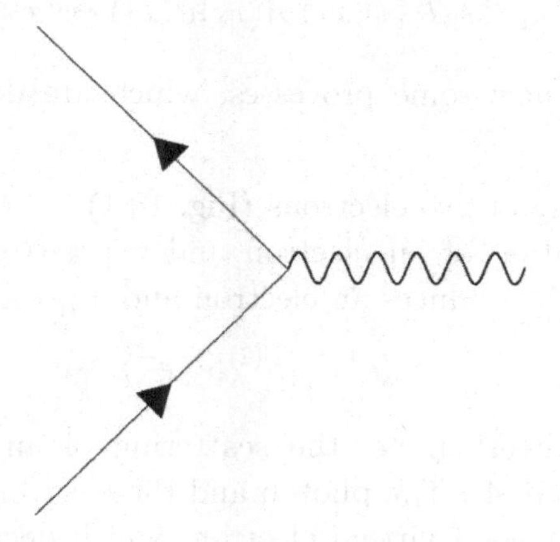

Fig. 10.2. The vertex describing the interaction of an electron and a photon.

The scattering processes are described by the S-matrix, which is given by the sum:

$$S = \sum_{n=0}^{\infty} \frac{1}{n!}(-ie)^n \int_{-\infty}^{\infty} d^4x_1 \cdots \int_{-\infty}^{\infty} d^4x_n \cdot T\left\{(\bar{\psi}(x_1)\cdot \not{A}(x_1)\right.$$
$$\left. \cdot\, \psi(x_1)) \cdots (\bar{\psi}(x_n)\cdot \not{A}(x_n)\cdot \psi(x_n))\right\}.$$

The vertex does not contribute to the S-matrix, due to energy-momentum conservation. The first term, which contributes, is of

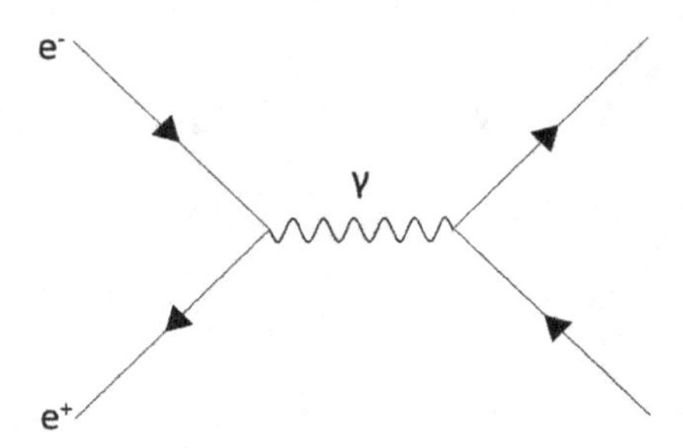

Fig. 10.3. Annihilation of an electron and a positron into a virtual photon, which decays into an electron and a positron.

second order:

$$S^{(2)} \propto e^2 \iint d^4x_1 d^4x_2 T\left\{\bar{\psi}(x_1) \not{A}(x_1)\psi(x_1) \cdot \bar{\psi}(x_2) \not{A}(x_2)\psi(x_2)\right\}.$$

We consider now some processes, which are described by this term:

(a) The scattering of two electrons (Fig. 10.1).
(b) The annihilation of an electron and a positron into a virtual photon, which produces an electron and a positron (Fig. 10.3):

$$\langle e^+e^-|S^{(2)}|e^+e^-\rangle.$$

(c) Compton scattering, i.e. the scattering of an electron and a photon (Fig. 10.4). The photon and the electron interact — the electron becomes a virtual electron, which decays immediately into a real electron and a photon. The relevant matrix element is given by:

$$\langle e^-\gamma|S^{(2)}|e^-\gamma\rangle.$$

(d) The annihilation of an electron and a positron into two photons (see Fig. 10.5). It is given by the matrix element:

$$\langle e^+e^-|S^{(2)}|\gamma\gamma\rangle.$$

The electromagnetic coupling constant e is given by the fine-structure constant, which is a function of the energy, due to the effects of vacuum polarization. The vacuum in quantum electrodynamics is

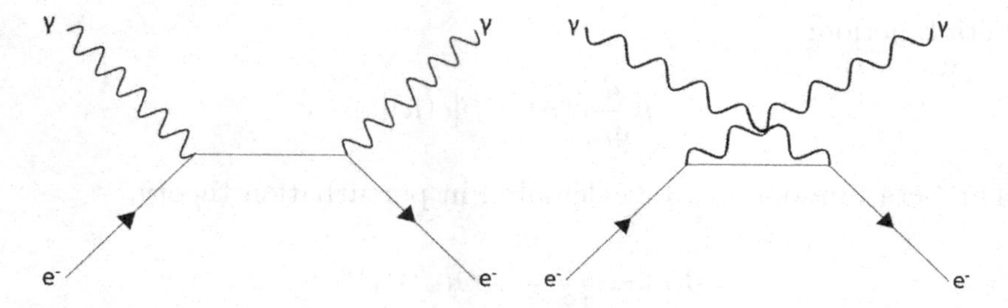

Fig. 10.4. The two Feynman diagrams for the Compton scattering.

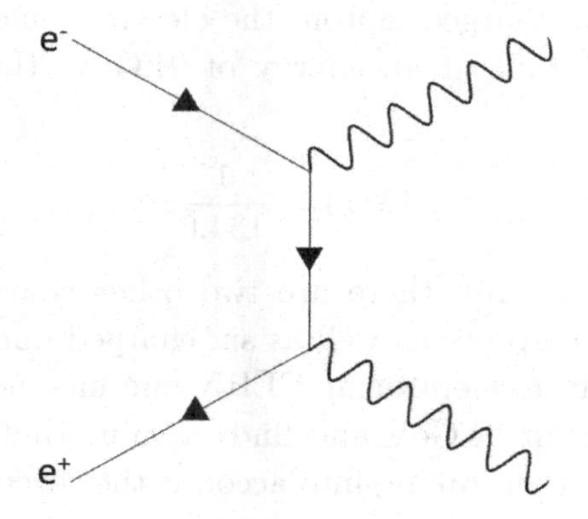

Fig. 10.5. The annihilation of electron and positron into two photons.

not empty, but contains virtual electron-positron pairs. A free electron repels the virtual electrons, but attracts the virtual positrons. Thus the electric charge of the electron is reduced. This reduction is a function of the energy or the corresponding distance. If the energy is increased, the coupling constant increases logarithmically.

We consider two different energies and calculate the change of the fine-structure constant:

$$\mu > \tau,$$

$$\Rightarrow \alpha(\mu^2) \cong \frac{\alpha(\tau^2)}{1 - \frac{\alpha(\tau^2)}{3\pi} \ln\left(\frac{\mu^2}{\tau^2}\right)}.$$

The variation of the coupling constant is described by the Callan-Symanzik equation. The logarithmic variation of e is given by the beta function:

$$\mu \frac{d}{d\mu} e(\mu) = \beta[e(\mu)].$$

The beta function can be calculated in perturbation theory:

$$\beta(e) = \frac{e^3}{12\pi^2} + O(e^5) + \cdots$$

The coupling constant e increases with increasing energy. If there would be only one charged lepton, the electron, one obtains for the fine-structure constant at an energy of 91 GeV, the mass of the Z boson:

$$\alpha(M_Z) \cong \frac{1}{134.6}.$$

But besides the electron there are two other charged leptons, the muon and the tau lepton, as well as six charged quarks.

With the LEP accelerator at CERN one has measured the fine-structure constant at 91 GeV and finds a value that agrees with the theoretical calculation, taking into account the three charged leptons and the six quarks:

$$\alpha(M_Z) \cong \frac{1}{(128.87 \pm 0.12)}.$$

The anomalous magnetic moment of the electron can be calculated in quantum electrodynamics. In general, the magnetic moment of a fermion is given by the mass of the fermion, the gyromagnetic factor g, the charge q and the spin S:

$$\vec{\mu} = g \cdot \left(\frac{q}{2m}\right) \vec{S}.$$

The gyromagnetic ratio of a free electron is $g = 2$. The difference between the actual value and $g = 2$ is the anomalous magnetic moment a:

$$a = \frac{g-2}{2}.$$

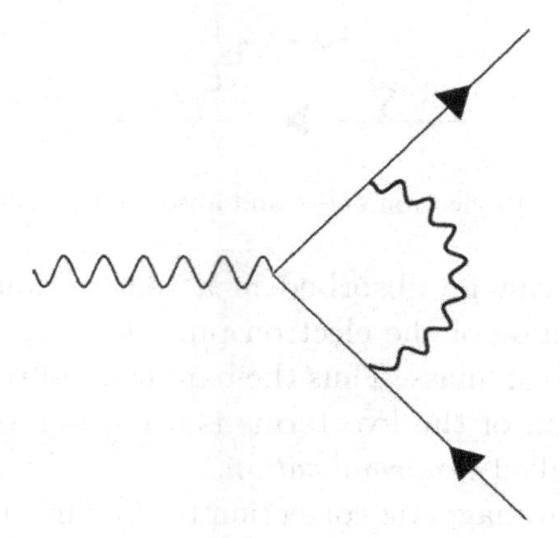

Fig. 10.6. The first correction to the anomalous magnetic moment of the electron is due to the exchange of a virtual photon.

In quantum electrodynamics, the anomalous magnetic moment of the electron can be calculated in perturbation theory. It is given by a power series in the fine-structure constant:

$$a_e = C_1 \cdot \frac{\alpha}{\pi} + C_2 \left(\frac{\alpha}{\pi}\right)^2 + C_3 \left(\frac{\alpha}{\pi}\right)^3 + C_4 \left(\frac{\alpha}{\pi}\right)^4 + \cdots$$

The first correction is due to the exchange of a virtual photon between the incoming and the outgoing electron (Fig. 10.6).

Thus far the first four coefficients have been calculated:

$$\begin{aligned} C_1 &= +0.500, \\ C_2 &\cong -0.328, \\ C_3 &\cong +1.181, \\ C_4 &\cong -1.914. \end{aligned}$$

To obtain the last coefficient, 891 Feynman diagrams had to be calculated.

The anomalous magnetic moment can be calculated, but other quantities are infinite, for example the electron mass. An electron can emit and reabsorb a virtual photon (Fig. 10.7) — this correction changes the mass of the electron by an infinite amount.

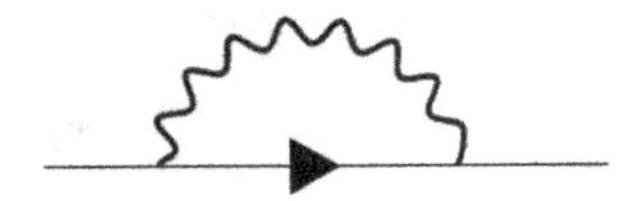

Fig. 10.7. An electron emits and absorbs a virtual photon.

This infinity can be absorbed in a redefinition of the electron mass. The bare mass of the electron plus the correction is given by the observed electron mass. Thus the bare mass of the electron is also infinite — the sum of the two terms is the observed electron mass. This process is called *renormalization.*

Also the electromagnetic correction to the fine-structure constant is infinite. This infinity is removed by setting the fine-structure constant equal to the observed fine-structure constant. There are no other infinities in quantum electrodynamics.

The muon is not stable, but decays into an electron and two neutrinos. It can be produced together with its antiparticle, if an electron and a positron annihilate to form a virtual photon, which decays into the muon and its antiparticle:

$$e^+e^- \rightarrow \mu^+\mu^-.$$

We calculate the differential cross section for this process, which is a function of the energy and the angle between the incoming electron and the produced muon, in the center-of-mass frame:

$$\frac{d\sigma}{d\Omega} = \frac{1}{16 \cdot \pi^2 \cdot E_{SP}} \cdot |M|^2.$$

The scattering amplitude is calculated in perturbation theory, using the corresponding Feynman diagrams — the lowest order diagram is shown in Fig. 10.8.

The scattering amplitude is given by the product of the photon propagator and of the two vertices:

$$M = \frac{e^2}{q^2}(\bar{v}(p')\gamma^\mu u(p)) \cdot (\bar{u}(k)\gamma_\mu v(k')).$$

The momentum of electron and positron is given by p and p'. The momenta of the muons are given by k and k'. The differential cross

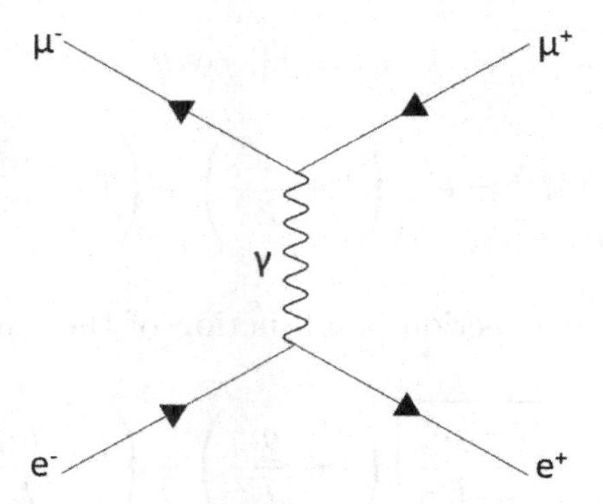

Fig. 10.8. The lowest order Feynman diagram of the annihilation of an electron and a positron into a muon pair.

section is given by the square of the scattering amplitude:

$$|M|^2 = \frac{e^4}{q^4}(\bar{v}(p')\gamma^\mu u(p) \cdot \bar{u}(p)\gamma_\mu v(p'))$$
$$\cdot (\bar{u}(k)\gamma_\mu v(k') \cdot \bar{v}(k')\gamma_\nu u(k)).$$

Here we average over the spins of the incoming fermions and sum over the spins of the outgoing fermions, using the relations:

$$\sum_s u^s(p) \cdot \bar{u}^s(p) = \not{p} + m,$$

$$\sum_s v^s(p) \cdot \bar{v}^s(p) = \not{p} - m.$$

$$\frac{1}{4}\sum_{Spin} |M|^2 = \frac{e^4}{4q^4}\mathrm{tr}[(\not{p}' - m_e)\gamma^\mu(\not{p} + m_e)\gamma^\nu]$$
$$\cdot\,\mathrm{tr}[(\not{k} + m_\mu)\gamma_\mu(\not{k}' - m_\mu)\gamma^\nu].$$

We calculate this in the center-of-mass frame:

$$p = (E, E_z), \quad p' = (E, -E_z),$$
$$k = (E, \vec{k}), \quad k' = (E, -\vec{k}),$$

$$|\vec{k}| = \sqrt{E^2 - m_\mu'^2}, \quad \vec{k} \cdot e_z = |\vec{k}| \cdot \cos\theta$$

$$\Rightarrow \frac{1}{4} \sum_{Spin} |M|^2 = e^4 \left[\left(1 + \frac{m_\mu^2}{E^2}\right) + \left(1 - \frac{m_\mu^2}{E^2}\right) \cos^2\theta \right].$$

The differential cross section is a function of the scattering angle:

$$\frac{d\sigma}{d\Omega} = \frac{\alpha^2}{4E_{SP}^2} \sqrt{1 - \frac{m_\mu^2}{E^2}} \left[\left(1 + \frac{m_\mu^2}{E^2}\right) + \left(1 - \frac{m_\mu^2}{E^2}\right) \cos^2\theta \right].$$

After the integration over the scattering angle we obtain the total cross section:

$$\sigma_{total} = \frac{4\pi\alpha^2}{3E_{SP}^2} \sqrt{1 - \frac{m_\mu^2}{E^2}} \left(1 + \frac{m_\mu^2}{2E^2}\right).$$

If the energy is much larger than the muon mass, we find:

$$\frac{d\sigma}{d\Omega} \to \frac{\alpha^2}{4E_{SP}^2}(1 + \cos^2\theta),$$

$$\sigma_{total} \to \frac{4\pi\alpha^2}{3E_{SP}^2}.$$

It is useful to calculate the cross section for the various orientations of the spins. An electron is a superposition of a left-handed and a right-handed electron:

$$e^- \to (e_L^-, e_R^-).$$

If an electron and a positron annihilate, a virtual photon with the spin 1 is created. The amplitude of the annihilation is zero if both fermions are left-handed or right-handed, since in this case the angular momentum is zero. Thus there are four different non-vanishing cross sections:

$$\frac{d\sigma}{d\Omega}(e_R^- e_L^+ \to \mu_R^- \mu_L^+) = \frac{\alpha^2}{4E_{SP}^2}(1+\cos\theta)^2,$$

$$\frac{d\sigma}{d\Omega}(e_R^- e_L^+ \to \mu_L^- \mu_R^+) = \frac{\alpha^2}{4E_{SP}^2}(1-\cos\theta)^2,$$

$$\frac{d\sigma}{d\Omega}(e_L^- e_R^+ \to \mu_R^- \mu_L^+) = \frac{\alpha^2}{4E_{SP}^2}(1-\cos\theta)^2,$$

$$\frac{d\sigma}{d\Omega}(e_L^- e_R^+ \to \mu_L^- \mu_R^+) = \frac{\alpha^2}{4E_{SP}^2}(1+\cos\theta)^2.$$

Chapter 11

Symmetry

If the physical properties of a system do not change under a transformation, one has a symmetry. There are discrete symmetries, for example the symmetry of a reflection of space, and continuous symmetries, for example the symmetry of the rotations of space or of the translation in time.

Each continuous symmetry implies a conserved quantity. For example, the symmetry of time translation is related to the conservation of energy. The symmetry of space rotation implies the conservation of the angular momentum.

There are symmetries, related to the properties of space and time, and internal symmetries, e.g. the isospin symmetry in nuclear physics or the SU(3) symmetry in particle physics.

Not only the symmetries are important, but also the breaking of symmetries. Often it is possible to describe the symmetry breaking by the symmetry. In this case one can obtain relations between the various parameters of the symmetry breaking. For example, if the SU(3) symmetry of particle physics is broken by an octet representation, one obtains relations for the masses of the particles in the corresponding representation.

There are different types of symmetry breaking, e.g. a spontaneous symmetry breaking, used to generate the masses of the weak bosons, and an explicit symmetry breaking, e.g. the breaking of the isospin symmetry in nuclear physics or the breaking of the SU(3) symmetry.

Space-time Symmetries

We consider an infinitesimal translation of space-time, described by a vector:

$$x^{\mu} \to x^{\mu} + a^{\mu}.$$

The change of the scalar field is given by the derivative of the field:

$$\phi(x) \to \phi(x+a) \cong \phi + \delta\phi = \phi(x) + a^{\mu}\partial_{\mu}\phi.$$

For the change of the Lagrange density we find:

$$\delta L = \frac{\partial L}{\partial \phi}\delta\phi + \frac{\partial L}{\partial(\partial_{\mu}\phi)}\delta(\partial_{\mu}\phi).$$

This change can be rewritten using the field equations:

$$\delta L = \partial_{\mu}\left(\frac{\partial L}{\partial(\partial_{\mu}\phi)}\delta\phi\right) = \partial_{\mu}\left(\frac{\partial L}{\partial\left(\partial_{\mu}\phi\right)}\partial_{\nu}\phi\right)a^{\nu},$$

$$\delta L = \partial_{\mu}L \cdot a^{\mu} = \delta_{\mu}^{\nu} \cdot \partial_{\mu}L \cdot a^{\nu},$$

$$\to \partial_{\mu}\left(\frac{\partial L}{\partial\left(\partial_{\mu}\phi\right)}\partial_{\nu}\phi - \delta_{\nu}^{\mu}L\right) = 0.$$

The tensor above is the energy-momentum tensor of the scalar field:

$$T^{\mu\nu} = \left(\frac{\partial L}{\partial\left(\partial_{\mu}\phi\right)}\partial^{\nu}\phi - g^{\mu\nu}L\right).$$

The time-time component of this tensor is the energy density:

$$T^{00} = \left(\frac{\partial L}{\partial\left(\partial_{0}\phi\right)}\partial^{0}\phi - L\right).$$

The three space-time components define the three momentum densities. One obtains the energy and the three momenta after integration:

$$E = \int d^{3}x \cdot T^{00},$$

$$P^{i} = \int d^{3}x \cdot T^{0i}.$$

Energy and momentum conservation follow from the symmetries of space and time translations, described by four parameters. The translations are a subgroup of the Lorentz group described by ten parameters: four parameters for the translations and six parameters for the proper Lorentz transformations.

We mention here two discrete symmetries. The parity transformation P is the reflection of the three space coordinates:

$$x \Rightarrow -x,$$

$$y \Rightarrow -y,$$

$$z \Rightarrow -z.$$

Parity is conserved in the classical mechanics, in the relativistic mechanics and in quantum mechanics. Also the electromagnetic and strong interactions conserve parity. But since 1956 it is known that the weak interactions violate parity. The parity violation was observed in the beta decay of cobalt-60 nuclei.

After the discovery of parity violation it was assumed that the product of parity and charge conjugation, i.e. the transition from particles to antiparticles, is conserved. This symmetry is called *CP symmetry*. But in 1964 it was observed, that the CP symmetry is also violated in the decay of K mesons. However the violation of the CP symmetry is much smaller than the violation of parity.

Isospin

The strong interactions do not discriminate between protons and neutrons. Both particles are considered to be two different components of the nucleon. They are the two states of a doublet of a new symmetry, the isospin symmetry, introduced in 1932 by Werner Heisenberg. This symmetry is not an exact symmetry — it is violated by the electromagnetic interaction. But the symmetry breaking is relatively small, about 1%.

The symmetry group of isospin is the group SU(2), the group of two-dimensional unitary matrices. This group has infinitely many representations: singlets, doublets, triplets, etc. The three generators of the isospin group are given by the three Pauli matrices:

$$\sigma_1 = \begin{pmatrix} 0 & 1 \\ 1 & 0 \end{pmatrix}, \quad \sigma_2 = \begin{pmatrix} 0 & -i \\ i & 0 \end{pmatrix}, \quad \sigma_3 = \begin{pmatrix} 1 & 0 \\ 0 & -1 \end{pmatrix}.$$

The Pauli matrices have the following properties:

$$(\sigma_k)^2 = 1,$$

$$\sigma_i \sigma_k + \sigma_k \sigma_i = 0 \quad i \neq k,$$

$$\left[\left(\frac{1}{2}\sigma_i \right), \left(\frac{1}{2}\sigma_j \right) \right] = i\varepsilon_{ijk} \left(\frac{1}{2}\sigma_k \right).$$

The third Pauli matrix is diagonal. The rank of a unitary group is given by the number of diagonal matrices — the group SU(2) has rank one. The commutation relations of the Pauli matrices define the Lie algebra of the group SU(2):

$$[T_i, T_j] = i\varepsilon_{ijk} T_k.$$

An isospin representation with isospin t has $(2t + 1)$ different states, which have different values for the third component of the isospin. The nucleon is an isospin doublet: $t = 1/2$:

$$N = \begin{pmatrix} p \\ n \end{pmatrix}.$$

The three π mesons form a triplet $(t = 1)$:

$$\pi = \begin{pmatrix} \pi^+ \\ \pi^0 \\ \pi^- \end{pmatrix}.$$

The four delta resonances are given by an isospin quadruplet $(t = 3/2)$:

$$\Delta = \begin{pmatrix} \Delta^{++} \\ \Delta^+ \\ \Delta^0 \\ \Delta^- \end{pmatrix}.$$

Also the atomic nuclei can be described by representations of the isospin. For example, a proton and a neutron form a deuteron, which is a singlet of isospin, given by the wave function:

$$D = \frac{1}{\sqrt{2}}|pn - np\rangle.$$

A carbon nucleus, also an isospin singlet, consists of six protons and six neutrons.

SU(3) Symmetry

After 1947 one discovered in the cosmic rays the four K mesons and the six hyperons: the neutral lambda baryon (mass ~1116 MeV), the three sigma baryons (mass ~1194 MeV) and the two chi baryons (mass ~1317 MeV).

The hyperons and the K mesons are *strange particles* — they carry a new quantum number, the strangeness S:

$$S(p) = S(n) = 0,$$
$$S(K^+) = S(K^0) = +1,$$
$$S(\Lambda) = S(\Sigma) = -1,$$
$$S(\Xi) = -2.$$

The sum of baryon number and strangeness is the hypercharge: $Y = B + S$. For the nucleons and the π mesons, the baryon number, the neutral isospin and the electric charge are related as follows:

$$Q = T_3 + \frac{1}{2}B.$$

But for the strange particles this relation is not valid and must be replaced by the Gell-Mann–Nishijima relation:

$$Q = T_3 + \frac{1}{2}(B + S) = T_3 + \frac{1}{2}Y,$$
$$Y = B + S.$$

In 1961 Murray Gell-Mann and Yuval Neeman introduced a new symmetry, described by the group SU(3), the group of the unitary 3×3 matrices. The isospin group is a subgroup of SU(3).

The group SU(3) has eight generators, among them the three generators of the isospin group and the hypercharge. These generators are described by the eight Gell-Mann matrices, which are analogous to the three Pauli matrices of the isospin group:

$$\lambda_1 = \begin{pmatrix} 0 & 1 & 0 \\ 1 & 0 & 0 \\ 0 & 0 & 0 \end{pmatrix}, \quad \lambda_2 = \begin{pmatrix} 0 & -i & 0 \\ i & 0 & 0 \\ 0 & 0 & 0 \end{pmatrix}, \quad \lambda_3 = \begin{pmatrix} 1 & 0 & 0 \\ 0 & -1 & 0 \\ 0 & 0 & 0 \end{pmatrix},$$

$$\lambda_4 = \begin{pmatrix} 0 & 0 & 1 \\ 0 & 0 & 0 \\ 1 & 0 & 0 \end{pmatrix}, \quad \lambda_5 = \begin{pmatrix} 0 & 0 & -i \\ 0 & 0 & 0 \\ i & 0 & 0 \end{pmatrix}, \quad \lambda_6 = \begin{pmatrix} 0 & 0 & 0 \\ 0 & 0 & 1 \\ 0 & 1 & 0 \end{pmatrix},$$

$$\lambda_7 = \begin{pmatrix} 0 & 0 & 0 \\ 0 & 0 & -i \\ i & 0 & 0 \end{pmatrix}, \quad \lambda_8 = \frac{1}{\sqrt{3}} \begin{pmatrix} 1 & 0 & 0 \\ 0 & 1 & 0 \\ 0 & 0 & -2 \end{pmatrix}.$$

Among them are two diagonal matrices, thus the group SU(3) has rank 2. The traces of these matrices vanish, and the trace of the product of two matrices is proportional to the unit matrix:

$$\mathrm{tr}\lambda_i = 0,$$

$$\mathrm{tr}(\lambda_i \lambda_j) = 2\delta_{ij}.$$

The commutators of the Gell-Mann matrices are described by the structure constants of the group SU(3):

$$\left[\left(\frac{1}{2}\lambda_i \right), \left(\frac{1}{2}\lambda_j \right) \right] = i f_{ijk} \left(\frac{1}{2}\lambda_k \right),$$

$$f_{123} = 1, \quad f_{147} = f_{246} = f_{257} = f_{345} = \frac{1}{2},$$

$$f_{156} = f_{367} = -\frac{1}{2},$$

$$f_{458} = f_{678} = \frac{\sqrt{3}}{2}.$$

The structure constants determine the Lie algebra of SU(3):

$$[T_i, T_j] = i f_{ijk}\, sT_k,$$
$$Y = \frac{2}{\sqrt{3}} T_8.$$

The irreducible representations of SU(3) are described by two integer numbers p and q. The number of states in a representation is a function of p and q:

$$Z = \frac{1}{2}(p+1)(q+1)(p+q+2).$$

In general a representation of SU(3) is a complex representation — the complex-conjugated representation is obtained by the exchange of p and q:

$$(p,q)^* = (q,p).$$

Examples: triplet ~(1,0), sextet ~(2,0), decuplet ~(3,0), octet ~(1,1).

The real representation (2,2) has 27 components, the complex representation (2,1) has 15 components. Products of irreducible representations transform as a sum of irreducible representations:

$$\bar{3} \otimes 3 = 1 \oplus 8,$$
$$3 \otimes 3 = 6 \oplus \bar{3},$$
$$3 \otimes 3 \otimes 3 = 1 \oplus 8 \oplus 8 \oplus 10,$$
$$8 \otimes 8 = 1 \oplus 8 \oplus 8 \oplus 10 \oplus \overline{10} \oplus 27,$$

The hadrons are described by irreducible representations of the group SU(3). The eight baryons are components of an octet representation (Fig. 11.1).

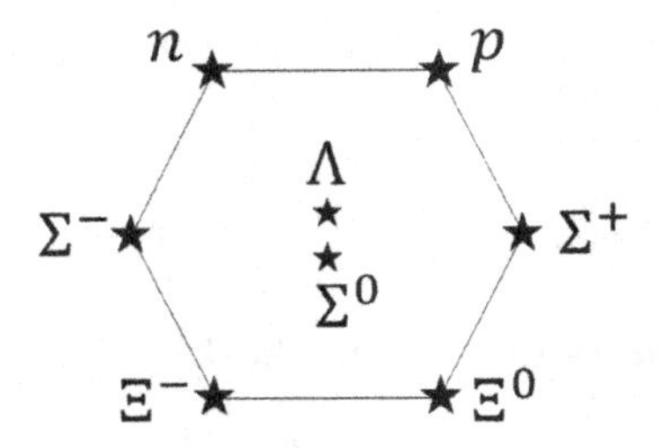

Fig. 11.1. The octet of the eight baryons.

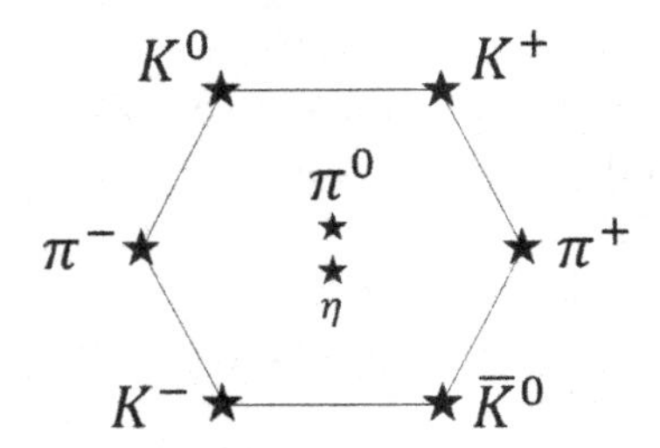

Fig. 11.2. The octet of the eight mesons.

Also the mesons form an octet, consisting of the three π mesons, the η meson and the four K mesons (Fig. 11.2).

The SU(3) symmetry is an approximate symmetry of the strong interactions. But the symmetry breaking is large, of the order of 20%. The isospin symmetry is violated by the electromagnetic interaction. But there is no interaction, which violates the SU(3) symmetry — the question arises, why is there such a large violation of SU(3)? Later we shall see that this violation can be understood in the quark model — it is related to the different masses of the quarks.

When the SU(3) symmetry was introduced, only 9 baryon resonances were known. There is an octet representation and a 10-representation, but no 9-dimensional representation. Gell-Mann assumed that the baryon resonances form a 10-representation. He introduced a tenth particle, which he called the omega minus. This particle (mass $\sim 1672\,\text{MeV}$) was discovered in 1964 at the Brookhaven National Laboratory (Fig. 11.3).

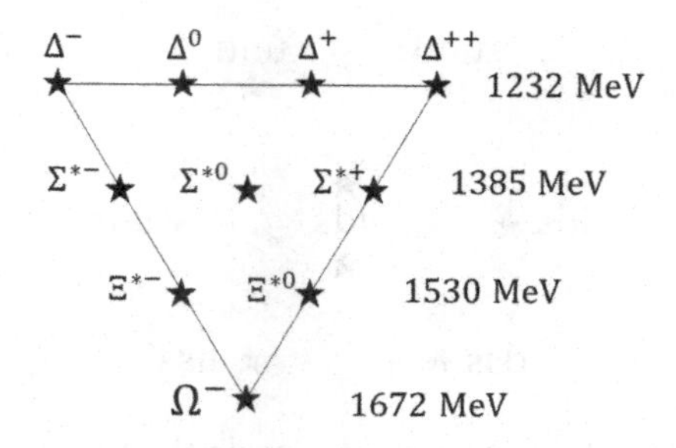

Fig. 11.3. The ten baryon resonances.

Quarks

The basic representation of isospin, the doublet representation, is present in nature since the nucleons form a doublet. The baryons and mesons are described by octets, decuplets and singlets of the SU(3) symmetry. But the basic representation, the triplet representation, is not observed in nature, neither is the sextet representation.

Today we know that the baryons and mesons are not elementary particles, as e.g. the electron, but bound states consisting of constituents which are SU(3) triplets, the *quarks.* Gell-Mann and George Zweig introduced the quark model in 1964. The term "quark" was found by Gell-Mann in the book *Finnegans Wake* by James Joyce.

Three quarks are needed, the up quarks, down quarks and strange quarks:

$$q = \begin{pmatrix} u \\ d \\ s \end{pmatrix}.$$

The quarks are fermions — they have spin 1/2. The electric charges of the quarks are peculiar:

$$e_Q = \begin{pmatrix} 2/3 \\ -1/3 \\ -1/3 \end{pmatrix} \cdot e.$$

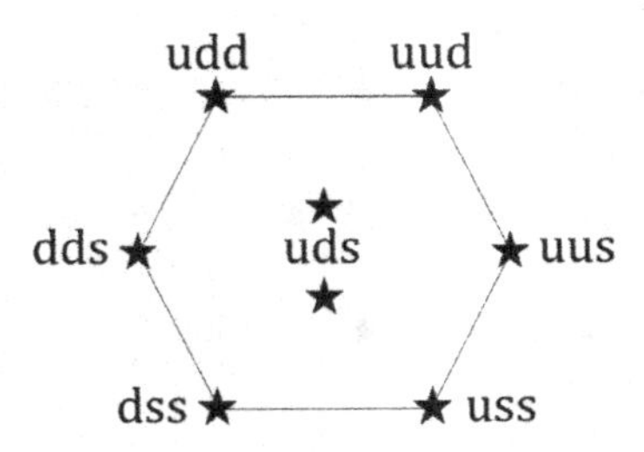

Fig. 11.4. The quarks inside the baryons.

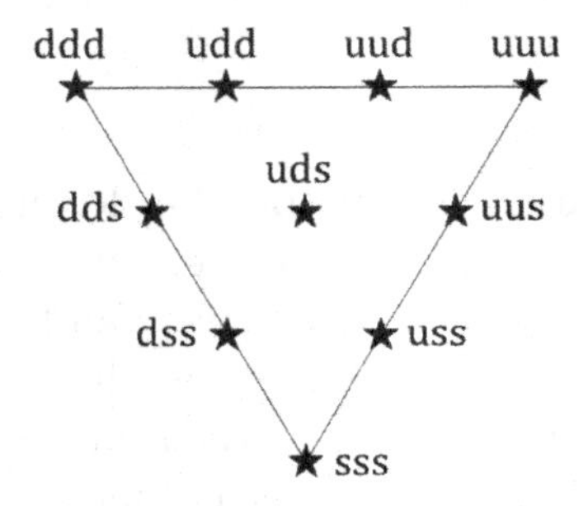

Fig. 11.5. The quarks inside the ten baryon resonances.

In the quark model, the group SU(3) is the group of the unitary transformations of the three quarks:

$$\begin{pmatrix} u \\ d \\ s \end{pmatrix}' = U \begin{pmatrix} u \\ d \\ s \end{pmatrix}.$$

The baryons are bound states of three quarks (Fig. 11.4).

Inside the lambda particle and the three sigma particles is one strange quark, the two chi particles have two strange quarks. The omega minus is a bound state of three strange quarks (Fig. 11.5).

The π mesons are bound states of one quark and an antiquark:

$$\pi^+ : (\bar{d}u),$$

$$\pi^0 : (\bar{u}u - \bar{d}d)/2,$$

$$\pi^- : (\bar{u}d).$$

Inside the K mesons is one strange quark or antiquark:

$$K^+ : (\bar{s}u), \quad K^0 : (\bar{s}d),$$
$$\bar{K}^0 : (\bar{d}s), \quad K^- : (\bar{u}s).$$

The quarks do not exist as free particles — they are permanently confined inside the hadrons. But they were observed indirectly in 1968 at the Stanford Linear Accelerator Center. In electron–proton collisions at high energy the electrons collided with the quarks and one could measure the momentum of a quark inside a proton. It turned out that the sum of the momenta of the quarks was only about 45% of the proton momentum. Later we shall see that the missing momentum is provided by the gluons — massless vector particles which bind the quarks inside a proton.

Although the quarks do not exist as free particles, it is possible to assign a mass to the quarks. If the quarks were massless, the SU(3) symmetry would be exact. The symmetry breaking is provided by the quark masses, thus information about the masses of the quarks can be obtained from the masses of the observed hadrons.

The masses of the up and down quarks are very small, about 5 MeV. The mass of the strange quark is much larger, about 100 MeV. Later we shall see that the quark masses depend on the relevant energy scale. The mass term of the quarks is given by the sum:

$$M_q = m_u\bar{u}u + m_d\bar{d}d + m_s\bar{s}s.$$

This term can be written in a different form:

$$M_q = m_0(\bar{u}u + \bar{d}d + \bar{s}s) + m_8(\bar{u}u + \bar{d}d - 2\bar{s}s)/\sqrt{3} + m_3(\bar{u}u - \bar{d}d).$$

Thus the SU(3) symmetry is broken by an SU(3) octet and an isospin doublet. The main part of the symmetry breaking transforms as an octet. Thus one obtains relations between the hyperon masses and the nucleon masses, which agree with the observed mass values.

Also the isospin symmetry is broken by the quark masses, since the up and down quarks have slightly different masses. The mass of the down quark is larger than the mass of the up quark. This explains why the neutron mass is larger than the proton mass. If the masses of the two quarks were the same, the proton mass would be slightly larger than the neutron mass, due to the electromagnetic self-energy of the proton.

Heavy Quarks

In 1974 a new meson, the **J**/ψ meson (mass $\sim$ 3.1 GeV), was discovered, which decays in particular into an electron and a positron. The lifetime of this meson was about a thousand times longer than expected. It turned out that the **J**/ψ meson was a bound state of a new quark, the charm quark, and its antiquark. The charm quark has the electric charge $2/3\,e$ and a mass of about 1100 MeV. One year later the mesons consisting of a charm quark and an up, down or strange antiquark, the D mesons, were discovered. They have masses of about 1.9 GeV.

Another heavy and long-lived meson was discovered in 1977 at the Fermi National Laboratory near Chicago, the upsilon meson (Υ). It has a lifetime of 1.21×10^{-20} s and a mass of about 9.46 GeV. Also the Υ meson is a bound state of a new quark, the bottom quark, and its antiquark. The bottom quark has the electric charge $-1/3\,e$ and a mass of about 4.3 GeV. Mesons that consist of one bottom quark and an up, down or strange antiquark are called B mesons. The masses of these mesons are about 5.3 GeV.

Finally, a sixth quark was discovered in 1995 also at the Fermi National Laboratory, the top quark, with the electric charge $2/3\,e$. It is the most massive of all observed elementary particles (mass $\sim$ 173 GeV).

The top quark interacts primarily by the strong interaction, but can only decay through the weak force. It decays to a W boson and either a bottom quark (most frequently), a strange quark, or, on the rarest of occasions, a down quark. The Standard Model predicts its mean lifetime to be roughly 5×10^{-25} s. This is about a twentieth of the timescale for the strong interactions. Therefore the top quark does not form hadrons before its decay.

Chapter 12

Gauge Theories

In quantum electrodynamics the field equations do not change, if the electron field is changed by a local phase transformation:

$$\psi(x) \to e^{i\lambda(x)}\psi(x).$$

Quantum electrodynamics is a gauge theory — the gauge group is the group of phase transformations U(1). Now we consider gauge theories with larger gauge groups, in particular with the groups SU(2) and SU(3).

We start with the Lagrange density of an SU(2) doublet of free complex scalar fields:

$$L = (\partial^\mu \phi^* \partial_\mu \phi) - m^2 \phi^* \phi,$$

$$\phi = \begin{pmatrix} \varphi_1 \\ \varphi_2 \end{pmatrix}.$$

The Lagrange density is invariant under global SU(2) transformations:

$$\phi \Rightarrow \phi' = U\phi.$$

Now we assume that the transformation is a local gauge transformation:

$$\phi \Rightarrow \phi' = U(x) \cdot \phi,$$

$$U = \begin{pmatrix} u_{11}(x) & u_{12}(x) \\ u_{21}(x) & u_{22}(x) \end{pmatrix}.$$

The matrix U can be written as an exponential, using the three Pauli matrices, multiplied with three parameters, which are functions of space and time:

$$U = \exp\left(i\alpha^k(x)\frac{\tau^k}{2}\right).$$

The derivative in the Lagrange density must be replaced by a covariant derivative which contains a vector field — the corresponding gauge field:

$$\partial_\mu\phi \Rightarrow D_\mu\phi = \left(\partial_\mu - ig\vec{A}_\mu\right)\phi.$$

The vector field is the sum of three gauge fields, which are multiplied with the corresponding Pauli matrix:

$$\vec{A}_\mu = A_\mu^k \cdot \frac{\tau_k}{2}.$$

The unitary matrix U and its derivative determine the change of the gauge field in a gauge transformation:

$$\vec{A}_\mu \Rightarrow \vec{A}'_\mu = U\vec{A}_\mu U^\dagger + (i/g)\cdot(\partial_\mu U)\cdot U^\dagger,$$
$$D'_\mu\phi' = U \cdot D_\mu\phi.$$

Here is the Lagrange density of the scalar field and the gauge field:

$$L = (D^\mu\phi^* D_\mu\phi) - m^2\phi^*\phi - \frac{1}{4}G^i_{\mu\nu}G^{\mu\nu}_i.$$

The field strengths of the three gauge fields are functions of derivatives of the gauge fields and of squares of the gauge fields multiplied with the gauge coupling:

$$G^i_{\mu\nu} = \partial_\mu A^i_\nu - \partial_\nu A^i_\mu + g\varepsilon^{ijk}A^j_\mu A^k_\nu.$$

The field strength can also be written as a tensor, which includes the commutator:

$$\vec{G}_{\mu\nu} = \partial_\mu\vec{A}_\nu - \partial_\nu\vec{A}_\mu + ig\left[\vec{A}_\mu, \vec{A}_\nu\right].$$

In quantum electrodynamics, which has only one gauge field, the third term in the field strength is absent.

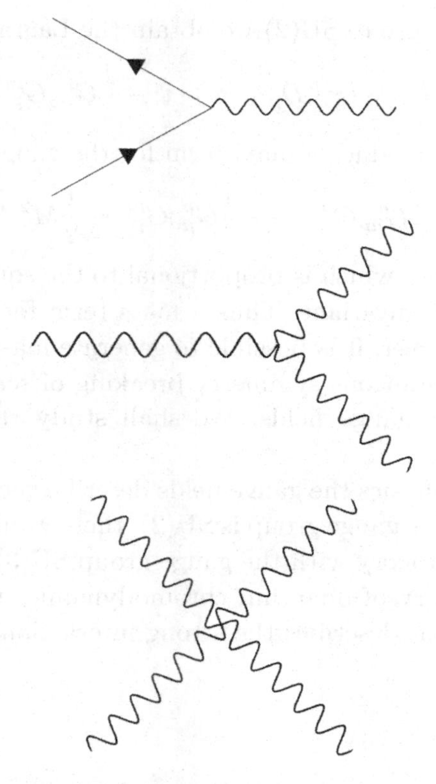

Fig. 12.1. The interactions of the gauge fields with the scalar fields and with themselves.

The Lagrange density contains terms that describe the interaction of the gauge fields with the scalar fields, and terms that describe the interaction of three and four gauge fields with themselves (Fig. 12.1).

In the same way a gauge theory with spinor fields can be constructed. We start with the Lagrange density of an SU(2) doublet of spinor fields:

$$L = \bar{\psi} \cdot (i\gamma^{\mu}\partial_{\mu} - m) \cdot \psi,$$
$$\psi = \begin{pmatrix} \psi_1 \\ \psi_2 \end{pmatrix}.$$

After gauging the group SU(2) we obtain the Lagrange density:

$$L = \bar{\psi} \cdot (i\gamma^\mu D_\mu - m) \cdot \psi - \frac{1}{4} G^i_{\mu\nu} G^{\mu\nu}_i .$$

It is possible to introduce a mass term for the gauge fields:

$$L = -\frac{1}{4} G^i_{\mu\nu} G^{\mu\nu}_i \rightarrow -\frac{1}{4} G^i_{\mu\nu} G^{\mu\nu}_i - \frac{1}{2} M^2 A_\mu A^\mu .$$

But this mass term, which is proportional to the square of the gauge field, is not gauge invariant. Thus a mass term for the gauge fields is forbidden. However, it is possible to generate masses for the gauge fields by the spontaneous symmetry breaking of scalar fields, which interact with the gauge fields. We shall study this possibility in Chapter 14.

In quantum physics the gauge fields describe vector particles, the gauge bosons. If the gauge group is SU(2), there would be three gauge bosons. A gauge theory with the gauge group SU(3) has eight gauge bosons. This theory of quantum chromodynamics will be discussed in Chapter 13 — it describes the strong interactions.

Chapter 13

Quantum Chromodynamics

The Colors of the Quarks

The omega resonance consists of three strange quarks. The angular momenta of the quarks are zero and the three spins are aligned. The wave function is symmetric under the exchange of two quarks, since it is the ground state.

However, according to the Pauli principle the wave function must be antisymmetric! To solve this problem, Gell-Mann and the author introduced in 1971 a new quantum number, the *color* of the quarks. There is a red up quark, a green up quark and a blue up quark, etc. The transformations of the colors are described by the color group SU(3), which is an exact symmetry. The quarks are color triplets:

$$\begin{pmatrix} u_r \\ u_g \\ u_b \end{pmatrix} \Rightarrow U \begin{pmatrix} u_r \\ u_g \\ u_b \end{pmatrix}.$$

The observed hadrons are singlets of the color group. The following terms are the simplest color singlets:

$$\text{Meson} \Rightarrow (\bar{q}q),$$

$$\text{Baryon} \Rightarrow (qqq),$$

$$\text{Antibaryon} \Rightarrow (\bar{q}\bar{q}\bar{q}).$$

As an example we consider the wave function of the positively charged pion:

$$|\pi^+\rangle = \frac{1}{\sqrt{3}}|(\bar{d}_r u_r + \bar{d}_g u_g + \bar{d}_b u_b)\rangle.$$

The wave functions of the baryons are antisymmetric in color, e.g.:

$$\begin{aligned}\Omega^- = |sss\rangle &\to \frac{1}{\sqrt{6}}(|s_r s_g s_b\rangle - |s_g s_r s_b\rangle + \cdots)\\ &= \frac{1}{\sqrt{6}}\sum_{i,j,k}\varepsilon_{ijk}|s_i s_j s_k\rangle.\end{aligned}$$

Thus the wave function is antisymmetric with respect to the exchange of two quarks, as required by the Pauli principle.

It was known that in the quark model there is a problem to understand the radiative decay of the neutral pion. The quark and the antiquark annihilate into two photons (Fig. 13.1). One obtains a decay rate which is a factor nine less that the observed rate.

The decay rate is a function of the pion decay constant and the number of colors:

$$\Gamma(\pi^0 \to 2\gamma) = \left(\frac{n_c}{3}\right)^2 \frac{\alpha^2}{64\pi^3} \cdot \frac{m_\pi^3}{f_\pi^2}.$$

This decay rate agrees well with the observed decay rate, if there are three colors. If there were no colors, it would be a factor nine smaller!

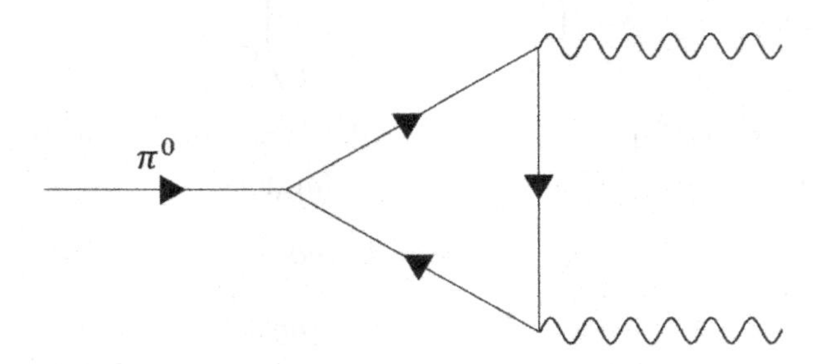

Fig. 13.1. The triangle diagram, describing the decay of the neutral pion into two photons.

Thus the neutral pion decay indicated that the color quantum number exists.

Another hint towards the color quantum number came from the experiments on electron–positron annihilation. If an electron and a positron annihilate, a muon pair or a quark–antiquark pair are produced. Let us consider the ratio R, the ratio of the cross sections for hadron production and muon pair production:

$$R = \frac{\sigma(e^+e^- \Rightarrow \text{Hadrons})}{\sigma(e^+e^- \Rightarrow \mu^+\mu^-)}.$$

In the quark model this ratio is given by the sum of the squares of the electric charges of the quarks:

$$R = \left(\frac{2}{3}\right)^2 + \left(-\frac{1}{3}\right)^2 + \left(-\frac{1}{3}\right)^2 = \frac{2}{3}.$$

According to the experiments this ratio is about 2.1 at the energy of about 3 GeV, more than three times larger. If the quarks have color, R should be 2, in good agreement with the experimental value.

Gauge Theory of Colored Quarks

In 1972 Murray Gell-Mann and the author used the color group as a gauge group of a gauge theory, which later was called the theory of quantum chromodynamics (QCD).

This theory is very similar to quantum electrodynamics. The electric charge is replaced by the eight color charges; the gauge group U(1) is replaced by the color group SU(3). The gauge boson of QED, the photon, is replaced by the eight gauge bosons of QCD, the gluons. Each quark is described by a triplet of colored quarks:

$$q = \begin{pmatrix} q_r \\ q_g \\ q_b \end{pmatrix}.$$

A color transformation is described by a unitary matrix U, which depends on space and time:

$$q \Rightarrow q' = U(\vec{x}, t) \cdot q.$$

As in QED we introduce a gauge field, which appears in the covariant derivative:

$$D_\mu q = (\partial_\mu + iA_\mu)q.$$

The gauge field is a color octet, i.e. the sum of eight vector fields multiplied with the corresponding Gell-Mann matrices:

$$A_\mu = \frac{1}{2}\sum_{a=1}^{8} A_\mu^a \lambda_a.$$

The gauge field changes under a color transformation:

$$A_\mu \Rightarrow A'_\mu = UA_\mu U^\dagger + (i/g)\cdot(\partial_\mu U)\cdot U^\dagger,$$
$$D'_\mu q' = U \cdot D_\mu q.$$

The field strengths of the gluons are a color octet:

$$G_{\mu\nu}^a = \partial_\mu A_\nu^a - \partial_\nu A_\mu^a + gf^{abc}A_\mu^b A_\nu^c,$$
$$G_{\mu\nu} = \partial_\mu A_\nu - \partial_\nu A_\mu - ig[A_\mu, A_\nu].$$

The Lagrange density of QCD is a function of the quark fields and of the gluon field strengths:

$$L = \bar{q}\left(i\gamma_\mu(\partial^\mu + ig\frac{\lambda^a}{2}A_a^\mu) - m\right)q - \frac{1}{4}G_{\mu\nu}^a G_a^{\mu\nu}.$$

In QED two electrons repel each other, an electron and a positron attract each other. In QCD the force depends on the colors of the quarks. A quark and an antiquark can attract each other or repel each other, depending on the colors. The product of a triplet and an anti-triplet is a singlet and an octet:

$$\bar{3} \otimes 3 = 1 \oplus 8.$$

The quark and the antiquark attract each other in a singlet, but repel each other in the octet.

Two quarks can form a color sextet or a color anti-triplet:

$$3 \otimes 3 = 6 \oplus \bar{3}.$$

If the quarks are in a sextet, they repel each other; in the anti-triplet they attract each other.

The Lagrange density of QCD contains the square of the gluon field strength. Since in the field strength the square of the gluon potential also appears, there is a direct interaction of three or four gluons. Such an interaction is not present in QED. This direct interaction of the gluons is important, since it leads to the asymptotic freedom of QCD.

Asymptotic Freedom

The interaction of the gluons with themselves has an interesting consequence — the coupling constant of QCD decreases if the energy is increased. The energy dependence of the coupling constant is described by a differential equation:

$$\mu \frac{d}{d\mu} g(\mu) = \beta(g).$$

Here appears the beta function, which is a power series of the coupling constant g starting with the third power. Thus in the lowest order one finds:

$$\mu \frac{d}{d\mu} g(\mu) \cong -\frac{1}{16\pi^2}\left(11 - \frac{2}{3} n_f\right) \cdot g^3(\mu).$$

Here $n(f)$ is the number of different quarks. The sign is negative, if the number of quarks is fewer than 16. In this case the coupling constant decreases logarithmically with the energy:

$$\frac{g^2}{4\pi} = \alpha_s(\mu^2) \cong \frac{\alpha_s(\mu_0^2)}{1 + \frac{1}{4\pi} \cdot \left(11 - \frac{2}{3} n_f\right) \cdot \alpha_s(\mu_0^2) \cdot \log\left(\frac{\mu^2}{\mu_0^2}\right)}.$$

Introducing the QCD scale parameter Λ, one finds:

$$\alpha_s(\mu^2) \cong \frac{4\pi}{\left(11 - \frac{2}{3} n_f\right) \log\left(\frac{\mu^2}{\Lambda^2}\right)}.$$

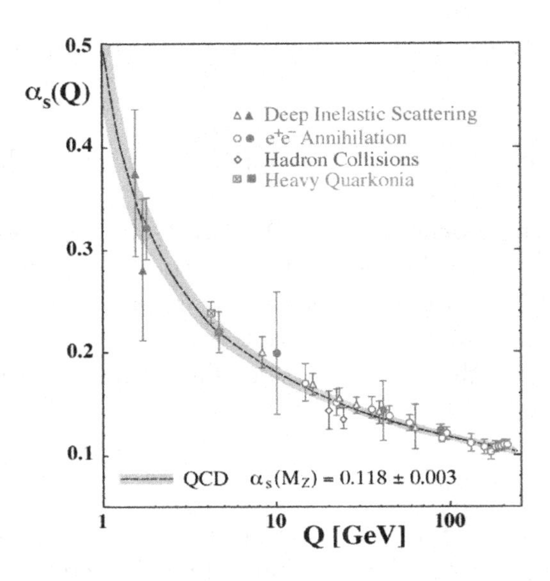

Fig. 13.2. The QCD coupling decreases with increasing energy.

The QCD scale parameter describes the energy scale at which the coupling constant becomes larger than one. It must be determined by experiment. The experiments give:

$$\Lambda = 213^{+38}_{-35}\ \text{MeV}.$$

The energy dependence of the QCD coupling has been measured at SLAC, at DESY and at CERN (Fig. 13.2).

The experimental data agree well with the theoretical predictions. With the LEP accelerator at CERN one has determined the QCD coupling at the mass of the Z boson with high precision:

$$\alpha_s(M_Z) = 0.1184 \pm 0.0007.$$

Since the QCD coupling increases with decreasing energy, the forces between the quarks at large distances become strong. Perturbation theory cannot be used in this region. But one can get information about the details of the forces in the lattice gauge theory — space and time are considered to be discrete quantities and can be described by a lattice.

In quantum electrodynamics the attractive force between an electron and a positron decreases, if the two particles are moved away from each other. In QCD the force between a quark and an antiquark does not decrease, but remains constant. The gluon field lines attract each other due to the self-interaction of the gluons.

At large distances the quark and the antiquark are connected by a string of parallel gluon field lines. The potential has a term linear in the distance between quark and antiquark:

$$V(r) \approx -\frac{4}{3} \cdot \frac{\alpha_s(r)}{r} + k \cdot r.$$

The constant k can be calculated in the lattice gauge theory. This potential describes rather well the spectrum of the bound states of heavy quarks.

In the presence of light quarks the field lines between a heavy quark and antiquark do not become parallel. If a heavy quark and antiquark are moved away from each other, the gluons produce light quark–antiquark pairs. The heavy quark attracts a light antiquark and forms a heavy meson (Fig. 13.3).

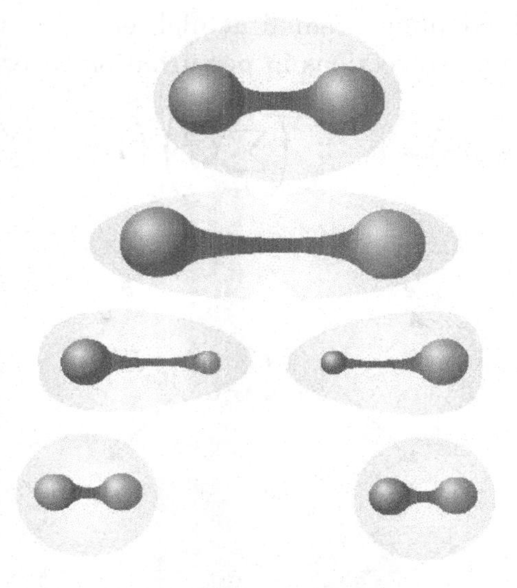

Fig. 13.3. A heavy quark and antiquark move away from each other — light quarks and antiquarks are produced.

The QCD coupling becomes large at low energies, i.e. at large distances. This indicates that the color force becomes strong enough that the quarks and gluons do not exist as free particles — they are confined inside the hadrons. But thus far a proof of the confinement does not exist.

Electron–Positron Annihilation

The cross section for the production of a muon pair in electron–positron annihilation is given by the fine-structure constant:

$$\sigma_0 = \frac{4\pi\alpha^2}{3s}, \qquad s = E_{cm}^2.$$

In QCD the production of hadrons is described by the production of quarks and antiquarks. If the quark–gluon interaction is neglected, the cross section is given by the sum of the squares of the quark charges and the number of colors:

$$\sigma(e^+e^- \to Hadrons) = \sigma_0 \cdot 3 \cdot \left(\sum_i Q_i^2\right).$$

Since the QCD coupling is small at high energies, it is possible to calculate the gluon corrections in perturbation theory (Fig. 13.4):

$$\sigma(e^+e^- \to Hadrons) = \sigma_0 \cdot 3 \cdot \left(\sum_i Q_i^2\right)\left(1 + \frac{\alpha_s\left(\sqrt{s}\right)}{\pi} + O\left(\alpha_s^2\right)\right).$$

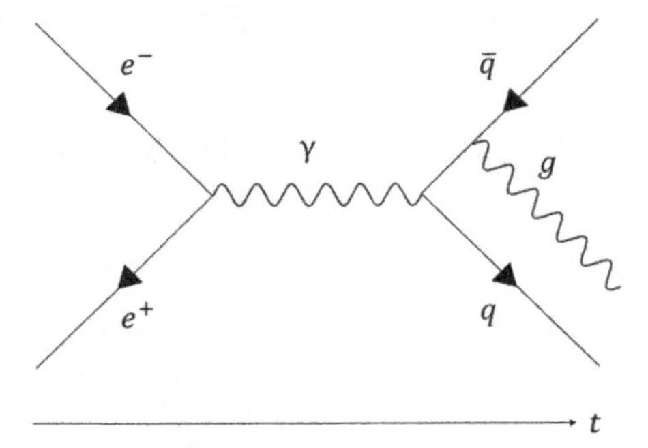

Fig. 13.4. Emission of a gluon in the annihilation of electron and positron.

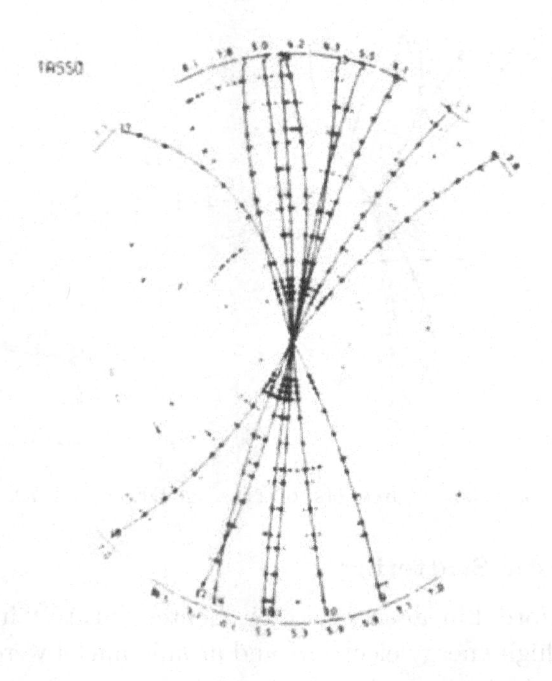

Fig. 13.5. Quark jets, observed at DESY in 1978.

The quarks do not exist as free particles, however they can be observed at high energies by observing jets of particles, as predicted by Richard Feynman in 1975. In the annihilation of an electron and a positron a quark and an antiquark with high energies are produced. The two quarks fragment and produce two jets of particles. The sum of the energies of the hadrons in a jet is equal to the energy of the quark. In 1978 the quark jets were observed with the PETRA accelerator at DESY in Hamburg (Fig. 13.5).

Sometimes a high energy gluon is produced together with the quark and antiquark:

$$e^+e^- \longrightarrow q\bar{q}g.$$

The gluon will also produce a particle jet — thus one should observe three jets. They were discovered in 1979 at DESY (Fig. 13.6).

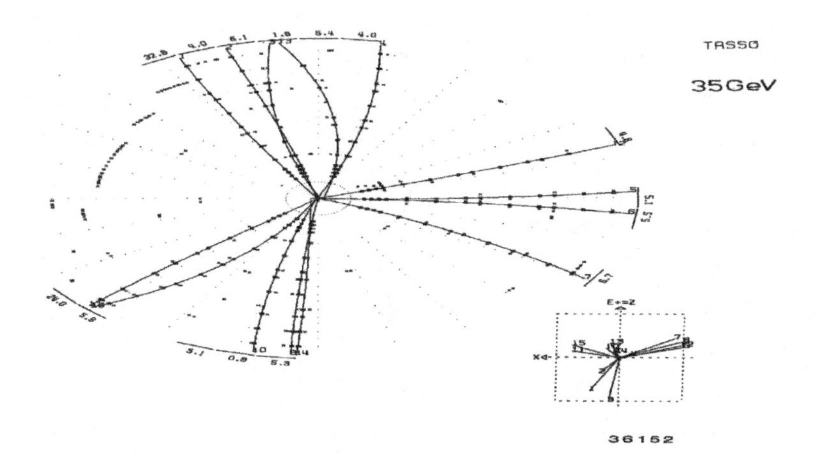

Fig. 13.6. Three jets, observed at DESY in 1979.

Deep-Inelastic Scattering

At the Stanford Linear Accelerator Center (SLAC) in California, collisions of high energy electrons and atomic nuclei were studied for the first time in 1968:

$$e^- + \text{nucleus} \Rightarrow e^- + X.$$

Let us consider the scattering of an electron and a proton. The differential cross section is given by the two structure functions:

$$\frac{d^2\sigma}{dEd\Omega} = \frac{e^4}{4E^2 \sin^4\left(\frac{\theta}{2}\right)} \left\{ \cos^2\left(\frac{\theta}{2}\right) \cdot W_2(\nu, Q^2) \right.$$
$$\left. + 2W_1(\nu, Q^2) \cdot \sin^2\left(\frac{\theta}{2}\right) \right\}.$$

Here the energy of the scattered electron is denoted by E, the angle is the scattering angle. The structure functions depend on two variables, the virtual mass of the photon, emitted from the electron, and the energy transfer from the electron to the nucleus:

$$Q^2 = -q^2, \quad \nu = \frac{P \cdot q}{M}.$$

At high energies one observed that the structure functions depend only on the ratio, given by the scale parameter x:

$$x = Q^2/2M\nu,$$

$$W_1(q^2,\nu) \Rightarrow F_1(x),$$

$$\frac{\nu}{M}W_2(q^2,\nu) \Rightarrow F_2(x).$$

This scaling behavior is obtained, if one assumes, that the proton is a bound state of point-like quarks. The electron scatters on a quark inside the proton — at high energies the momentum of the quark is a fraction of the momentum of the proton:

$$p = \xi \cdot P,$$

$$0 < \xi < 1.$$

The scaling parameter x is the ratio of the quark momentum and the proton momentum:

$$\xi = \frac{Q^2}{2P \cdot q} = \frac{Q^2}{2M\nu} \rightarrow x.$$

Since the quarks are fermions, there is a relation between the two structure functions:

$$F_2(x) \cong 2xF_1(x).$$

The proton consists of two up quarks, one down quark and a cloud of quark–antiquark pairs and gluons. The distribution function of a quark inside the proton is given by $q(x)$, the distribution function of the gluons is denoted by $g(x)$. The measured structure function is a function of the distribution functions:

$$F_2(x) = x \cdot \left\{\frac{4}{9}(u(x) + \bar{u}(x)) + \frac{1}{9}(d(x) + \bar{d}(x)) + \frac{1}{9}(s(x) + \bar{s}(x))\right\}.$$

The distribution functions have been measured for the proton. Interesting is the following integral:

$$\int_0^1 x \cdot [u(x) + \bar{u}(x) + d(x) + \bar{d}(x) + s(x) + \bar{s}(x)] \cdot dx = \varepsilon.$$

This integral is the sum of the momenta of the quarks inside the proton. If the proton were to consist only of quarks and antiquarks, this integral should equal to one. However, the experiments indicate that it is only about 45%. Thus 55% of the proton momentum must come from the gluons:

$$\varepsilon \approx 0.45 \Rightarrow \text{quarks},$$

$$1 - \varepsilon \approx 0.55 \Rightarrow \text{gluons}.$$

If the gluon distribution function $g(x)$ is included, one finds:

$$\int_0^1 x \cdot [u(x) + \bar{u}(x) + d(x) + \bar{d}(x) + s(x) + \bar{s}(x) + g(x)] \cdot dx = 1.$$

Inside the proton are two up quarks, one down quark and many quark–antiquark pairs. The distribution functions must obey the following sum rules:

$$\int_0^1 dx(u - \bar{u}) = 2,$$

$$\int_0^1 dx(d - \bar{d}) = 1.$$

Violation of Scaling

At high energies, the QCD coupling is small, but not zero. Therefore, one expects that the scaling behavior of the structure functions is not exact. The quark distribution functions should change slowly, due to the emission of gluons:

$$u(x) \Rightarrow u(x, Q^2).$$

The violations of scaling were predicted by theorists and then observed in the experiments. The observed violations of the scaling give information about the strength of the QCD coupling. The QCD coupling parameter has been measured in particular at the mass of the Z boson (~91 GeV):

$$\alpha_s(M_Z) = 0.1184 \pm 0.0007.$$

The Masses of the Quarks

The masses of the quarks are free parameters of QCD, analogous to the electron mass in QED. They describe the breaking of the symmetries, e.g. the isospin symmetry or the SU(3) symmetry.

If the masses of the up, down and strange quarks were zero, not only the SU(3) symmetry would be an exact symmetry, but also the chiral symmetry SU(3) $\times$ SU(3), which describes the symmetry of the left-handed and of the right-handed quarks.

$$SU(3) \Rightarrow SU(3)_L \otimes SU(3)_R.$$

This group has 16 generators, which are the integrals of the eight vector currents and the eight axial-vector currents. If the three quarks are massless, these currents are conserved:

$$\partial^\mu j_{i\mu} = 0,$$

$$\partial^\mu j^5_{i\mu} = 0.$$

The chiral symmetry is not realized in the spectrum of the hadrons. If the three quark masses are zero, the masses of the eight pseudo-scalar mesons, the three pi mesons, the four kaons and the eta meson, vanish. They are "Goldstone" particles. If an axial charge is applied on a nucleon, one obtains a nucleon and a massless pion with zero momentum:

$$F^5_i = \int d^3x j^5_{i0}(x),$$

$$F^5_i|N\rangle = |N, \pi\rangle.$$

The masses of the light quarks, the up, down and strange quarks, can be calculated from the chiral symmetry. If the quark masses are small, the squares of the masses of the pseudo-scalar mesons are proportional to the quark mass. Thus one can determine the ratios of the quark masses. Let us assume that the isospin symmetry is exact:

$$m_u = m_d = m.$$

The pion mass and the quark mass m are related:

$$M_\pi^2 = C \cdot m \cdot \Lambda_c.$$

The dimensionless constant C can be calculated in QCD — it is about 15. Thus the quark mass m is about 5 MeV. Analogous relations can be obtained for the K mesons — the ratio of m and the strange quark mass can be determined.

In QCD the masses of the quarks are scale-dependent, i.e. they depend on the energy:

$$m \Rightarrow m(\mu).$$

The isospin symmetry is not exact, but also violated by the quark masses. If the energy is 2 GeV, one finds for the light quark masses:

$$m_u \approx 3\,\text{MeV},$$

$$m_d \approx 5\,\text{MeV},$$

$$m_s \approx 100\,\text{MeV}.$$

At the energy of 91 GeV, the mass of the Z boson, the quark masses are smaller:

$$m_u(M_Z) \approx 1.4\,\text{MeV},$$

$$m_d(M_Z) \approx 2.8\,\text{MeV},$$

$$m_s(M_Z) \approx 57\,\text{MeV}.$$

The masses of the up and down quarks are very small. If their masses vanish, one expects that there will be four conserved axial-vector currents — the three isospin triplet currents and the isospin singlet current:

$$\partial_\mu j_0^{\mu 5} = \partial_\mu(\bar{u}\gamma_\mu\gamma_5 u + \bar{d}\gamma_\mu\gamma_5 d) = 0.$$

Thus there should be four Goldstone bosons: the three massless pi mesons and a massless isospin singlet meson. However we observe only three light mesons. The reason for this phenomenon is the gluon anomaly.

Let us consider first the quantum electrodynamics. If the mass of the electron vanishes, the axial current should be conserved:

$$\partial_\mu j^{\mu 5} = \partial_\mu \bar{e}\gamma^\mu \gamma_5 e = 0.$$

But this is not correct — the divergence of this axial current is proportional to the square of the electromagnetic field strength:

$$\partial_\mu j^{\mu 5} = -\frac{e^2}{16\pi^2}\varepsilon^{\alpha\beta\mu\nu} F_{\alpha\beta} \cdot F_{\mu\nu}.$$

This result cannot be obtained from the field equations. It is due to the singularity of the triangle diagram (see Fig. 13.1), which generates the anomalous divergence. It also appears in the divergence of the neutral axial-current of the quarks and explains, why the neutral pi meson decays rapidly into two photons.

Now we replace the photons by gluons. In QCD there is also an anomalous divergence for the isospin singlet axial current:

$$\partial_\mu(\bar{u}\gamma^\mu\gamma^5 u + \bar{d}\gamma^\mu\gamma^5 d) = -\frac{g^2}{8\pi^2}\varepsilon^{\alpha\beta\mu\nu} G^a_{\alpha\beta} G^a_{\mu\nu}.$$

This current is not conserved if the masses of the quarks are zero. Thus there is no Goldstone particle associated with this current. If the masses of the up, down and strange quark vanish, there are eight Goldstone particles, i.e. eight massless pseudoscalar mesons. Due to the anomaly the ninth meson, the eta meson, has a mass of about 960 MeV.

The chiral symmetry is strongly broken for the charm quark and the bottom quark and cannot be used to determine their masses. They can be estimated by analyzing the spectrum of charmed or bottom hadrons:

$$m_c(m_c) \approx 1.2 - 1.3\,\text{GeV},$$

$$m_b(m_b) \approx 4.1 - 4.4\,\text{GeV}.$$

Here the masses are normalized at the corresponding quark mass.

The top quark is very heavy and decays immediately after its production, without forming bound states. It decays mostly into a bottom quark after emitting a charged weak boson. Now the top quark mass is much better known than the other quark masses:

$$m_t = 173.34 \pm 0.27(\text{stat.}) \pm 0.71(\text{syst.})\,\text{GeV}.$$

Chapter 14

Spontaneous Symmetry Breaking

Scalar Fields

Sometimes the field equations are symmetric, but the symmetry is broken by the ground state. This mechanism is called *spontaneous symmetry breaking*. In quantum field theory the spontaneous symmetry breaking can be arranged with scalar fields.

We consider a scalar field, which interacts with itself:

$$L = \frac{1}{2}(\partial^\mu \varphi \cdot \partial_\mu \varphi) - \frac{1}{2}m^2 \cdot \varphi^2 - \lambda \cdot \varphi^4.$$

The potential of the field is given by the sum of the mass term and the interaction term:

$$V(\varphi) = \frac{1}{2}m^2 \cdot \varphi^2 + \lambda \cdot \varphi^4.$$

At the minimum of the potential the field vanishes.

The system has a discrete symmetry R:

$$R : \varphi \Rightarrow -\varphi.$$

Also the ground state is symmetric:

$$R|0\rangle = |0\rangle.$$

Now we change the sign of the mass term:

$$
\begin{aligned}
m^2 &\Rightarrow -m^2, \\
L &= \frac{1}{2}(\partial^\mu \varphi \cdot \partial_\mu \varphi) + \frac{1}{2} m^2 \cdot \varphi^2 - \lambda \cdot \varphi^4, \\
V(\varphi) &= -\frac{1}{2} m^2 \cdot \varphi^2 + \lambda \cdot \varphi^4.
\end{aligned}
$$

The new potential still has the symmetry R, but now there are two minima:

$$\varphi_0 = \pm v = \pm m / 2\sqrt{\lambda}.$$

The constant v is called the vacuum expectation values of the field. Since there are two minima, the system has two different ground states with the vacuum expectation values v and $-v$. The ground states are not symmetric — the symmetry is broken spontaneously.

Now we consider a complex scalar field:

$$
\begin{aligned}
\phi &= \frac{1}{\sqrt{2}}(\varphi + i\psi), \\
L &= \partial^\mu \phi^* \cdot \partial_\mu \phi - m^2 \phi^* \phi - \lambda \cdot (\phi^* \phi)^2, \\
V &= m^2 \phi^* \phi + \lambda (\phi^* \phi)^2.
\end{aligned}
$$

If m^2 is negative, the symmetry is spontaneously broken. The potential is similar to a "Mexican hat." The minimum of the potential is given by the mass parameter, the coupling constant and a phase parameter (Fig. 14.1):

$$\phi(x) = \phi_0 = \sqrt{\frac{-m^2}{2\lambda}} e^{i\theta}, \quad 0 \leq \theta \leq 2\pi.$$

Now there are infinitely many ground states, since the phase parameter is arbitrary.

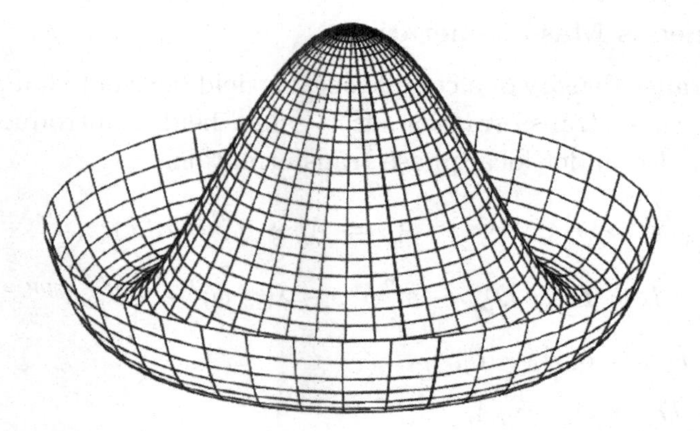

Fig. 14.1. The potential of a complex scalar field for negative m ("Mexican hat").

We choose the ground state in such a way that the phase parameter is zero:

$$\phi_0 = \sqrt{\frac{-m^2}{2\lambda}} = \frac{1}{\sqrt{2}} \cdot v.$$

The complex field is written as the sum of the real part and the imaginary part:

$$\phi(x) = \frac{1}{\sqrt{2}}[v + \sigma(x) + i\eta(x)].$$

The Lagrange density takes the following form:

$$L = \frac{1}{2}\partial^\mu\sigma \cdot \partial_\mu\sigma - \frac{1}{2}(2\lambda v^2) \cdot \sigma^2 + \frac{1}{2}\partial^\mu\eta \cdot \partial_\mu\eta$$
$$-\lambda \cdot v \cdot \sigma(\sigma^2 + \eta^2) - \frac{1}{4}\lambda(\sigma^2 + \eta^2)^2.$$

It describes a massive field and a massless field (*Goldstone field*):

$$\sigma : M = \sqrt{2\lambda} \cdot v,$$
$$\eta : M = 0.$$

Spontaneous Mass Generation

The Lagrange density of a complex scalar field does not change under a global phase transformation. If a gauge field is introduced, it is invariant also under local phase transformations:

$$\begin{aligned} \phi &\Rightarrow \phi \cdot e^{-i\alpha(x)}, \quad A_\mu \Rightarrow A_\mu + g^{-1}\partial_\mu \alpha(x), \\ L &= D^\mu \phi^* D_\mu \phi - m^2 \phi^* \phi - \lambda(\phi^* \phi)^2 - \frac{1}{4} F_{\mu\nu} F^{\mu\nu}, \\ F_{\mu\nu} &= \partial_\nu A_\mu - \partial_\mu A_\nu, \\ D_\mu &= \partial_\mu + igA_\mu. \end{aligned}$$

If the symmetry is spontaneously broken, we split up the complex scalar field into two real fields. The first field has a non-zero vacuum expectation value:

$$\phi(x) = \frac{1}{\sqrt{2}}[v + \sigma(x) + i\eta(x)].$$

Now the Lagrange density has a term proportional to the square of the gauge field:

$$g^2 v^2 A_\mu A^\mu.$$

This term is a mass term for the gauge boson — it is generated by a spontaneous symmetry breaking.

This mechanism is called the *Higgs mechanism*. It was introduced in 1964 by Robert Brout and Francois Englert, by Peter Higgs and by Gerald Guralnik, Carl Hagen and Tom Kibble.

A massless gauge boson has two degrees of freedom — it has either a right-handed polarization or a left-handed polarization. But a massive gauge boson has three degrees of freedom — it can also be longitudinally polarized. This additional degree of freedom is provided by the massless Goldstone field. The massless gauge boson absorbs the massless Goldstone particle and acquires a mass.

The number of degrees of freedom does not change. At the beginning there are four degrees of freedom, two for the massless gauge boson and two for the complex scalar field. After the symmetry breaking there are three degrees of freedom for the massive gauge boson and one degree of freedom for the massive scalar boson.

If the masses of the weak bosons are generated by the Higgs mechanism, there should exist a new scalar boson, the Higgs particle. The mass of this particle cannot be calculated, since it depends on the unknown parameter for the coupling of the scalar fields.

The Gauge Group SU(2)

Now we consider the spontaneous symmetry breaking for a gauge theory with the gauge group SU(2). It can be arranged with a doublet of complex scalar fields:

$$L = -\frac{1}{4} G_{\mu\nu}^i G_i^{\mu\nu} + \left| \left(\partial_\mu \phi + i \frac{g}{2} \tau^i B_\mu^i \phi \right) \right|^2 - m^2 \phi^* \phi - \lambda (\phi^* \phi)^2.$$

If m^2 is larger than zero, we have a system of massless gauge fields, interacting with a massive scalar field. If the symmetry is spontaneously broken ($m^2 < 0$), the scalar field has a non-vanishing vacuum expectation value. One can always arrange that the vacuum expectation value vanishes for the upper component of the scalar field:

$$\langle 0|\phi|0\rangle = \frac{1}{\sqrt{2}} \begin{pmatrix} 0 \\ v \end{pmatrix},$$

$$v = \sqrt{-m^2/\lambda}.$$

Now one has a mass term for the gauge bosons:

$$M^2 \propto (g^2/8) \cdot v^2 ((B_\mu^1)^2 + (B_\mu^2)^2 + (B_\mu^3)^2),$$

$$M = \frac{1}{2} g v.$$

Thus far the three gauge bosons have the same mass — the mass term is symmetric with respect to the group SU(2), since the scalar field is a doublet. In Chapter 15 we shall use this mechanism to generate the masses of the weak bosons in a theory, which unifies the weak interactions and the electromagnetic interactions — the electroweak gauge theory.

It was shown in 1971 by Gerard 't Hooft and Martinus Veltman that the electroweak gauge theory is renormalizable, if the gauge boson masses are introduced by a spontaneous symmetry breaking.

One can also break the symmetry using a triplet of real scalar fields:

$$\varphi = \begin{pmatrix} \phi_1 \\ \phi_2 \\ \phi_3 \end{pmatrix},$$

$$L = -\frac{1}{4} G^i_{\mu\nu} G^{\mu\nu}_i + \frac{1}{2} |(\partial_\mu \phi_i - g\varepsilon_{ijk} B^j_\mu \phi_k)|^2 - \frac{1}{2} m^2 (\phi_i \phi_i) - \lambda (\phi_i \phi_i)^2.$$

It is always possible to arrange that only the third scalar field has a non-vanishing vacuum expectation value:

$$\langle 0|\varphi|0\rangle = \begin{pmatrix} 0 \\ 0 \\ v \end{pmatrix}.$$

This vector does not change under rotations around the third axis, thus two gauge bosons become massive. The third gauge boson remains massless.

Chapter 15

Electroweak Interactions

The beta decay of the neutron, e.g. the decay into a proton, an electron and an antineutrino, is caused by the weak interactions. It can be described by an interaction given by the product of two weak currents multiplied with a coupling constant, the Fermi constant. This interaction was proposed by Enrico Fermi in 1934:

$$H \sim G_F J_\mu^* J_\mu,$$

$$J_\mu = \bar{n}\gamma_\mu p + \bar{e}\gamma_\mu \nu_e,$$

$$G_F \cong 1.1664 \times 10^{-5}\,\mathrm{GeV}^{-2}.$$

In 1956 it was discovered that the weak interactions violate parity. The electrons, emitted in the beta decay, are left-handed. The weak current is a left-handed current:

$$J_\mu = \bar{e}\gamma_\mu \left(\frac{1-\gamma_5}{2}\right) \nu_e + \bar{n}\gamma_\mu \left(\frac{1-\gamma_5}{2}\right) p.$$

This theory can only be an effective theory at low energies. The cross section for the scattering of a neutrino and an antineutrino increases at high energies in such a way, that the unitarity of quantum physics is violated.

It is possible to describe the weak interactions by the exchange of weak vector bosons, which must have a large mass. In the beta decay the neutron turns into a proton by emitting a virtual charged weak boson, which then produces an electron and an electron antineutrino.

The Lagrange density of this interaction is given by the product of the weak current and the field of the weak boson, multiplied with the weak coupling constant:

$$L^w \sim gW_\mu^- J^\mu + h.c.$$

In the beta decay the momentum transfer is very small — it is possible to replace the inverse propagator of the weak boson by the square of the mass of the weak boson. The Fermi constant is then given by the mass of the weak boson and the coupling constant:

$$\frac{G_F}{\sqrt{2}} \cong \frac{g^2}{8M_W^2}.$$

The weak bosons could be the gauge bosons of an electroweak theory, based on the gauge group SU(2) × U(1) — the weak interactions and the electromagnetic interactions are unified.

In this theory there are four gauge bosons: the two charged weak bosons, a neutral Z boson and the photon. The interaction, provided by the Z boson, is called the *neutral current interaction*. Thus besides the charged weak currents there is also a neutral current. An elastic scattering of neutrinos and protons is possible: neutrino + proton → neutrino + proton.

The neutral current was discovered at CERN in 1972. With the detector Gargamelle, the elastic scattering of neutrinos and atomic nuclei was observed.

One prediction of the electroweak theory is the violation of parity by the neutral current. In 1978 this parity violation was discovered at the Stanford Linear Accelerator Center in the United States in the experimental study of electron–nucleus collisions at high energy.

Electroweak Theory of the Electron and its Neutrino

We consider the electroweak gauge theory of electrons and neutrinos. The gauge group is SU(2) × U(1) — the group SU(2) describes the weak isospin.

The electron neutrino is a left-handed fermion. Together with the left-handed electron it forms a doublet of the weak isospin. The right-handed electron is a singlet of the weak isospin and does not interact weakly:

$$E_L = \begin{pmatrix} \nu_e \\ e^- \end{pmatrix}_L, \quad e_L^- = \left(\frac{1-\gamma_5}{2}\right) e^-, \quad e_R^- = \left(\frac{1+\gamma_5}{2}\right) e^-.$$

The electric charge Q is a superposition of the neutral generator of the weak isospin and the weak hypercharge Y:

$$Q = T_3 + \frac{1}{2}Y, \quad Y(e_L^-) = Y(\nu_e) = -1, \quad Y(e_R^-) = -2.$$

In the electroweak theory there are four gauge bosons: the three gauge bosons for the group SU(2) and one gauge boson for the group U(1). The Lagrange density of the gauge bosons is given by the square of the field strengths:

$$L = -\frac{1}{4}F_{\mu\nu}^i F_i^{\mu\nu} - \frac{1}{4}G_{\mu\nu}G^{\mu\nu},$$

$$F_{\mu\nu}^i = \partial_\mu W_\nu^i - \partial_\nu W_\mu^i + g\varepsilon_{ijk} W_\mu^j W_\nu^k,$$

$$G_{\mu\nu} = \partial_\mu B_\nu - \partial_\nu B_\mu.$$

In the doublet representation the three generators are described by the three Pauli matrices:

$$T_i = \frac{1}{2}\tau_i.$$

The leptons are described by a left-handed doublet and a right-handed singlet:

$$L^e = \bar{E}_L \left(i\partial\!\!\!/ - \frac{g}{2}\tau_i \cdot W\!\!\!\!/\,_i - \frac{g'}{2}B\!\!\!/ \right) E_L + \bar{e}_R(i\partial\!\!\!/ - g'B\!\!\!/)e_R.$$

The spontaneous symmetry breaking is arranged by a doublet of scalar complex fields:

$$\phi = \begin{pmatrix} \phi^+ \\ \phi^0 \end{pmatrix}, \quad Y(\phi) = +1,$$

$$L^H = (D^\mu\phi)^* (D_\mu\phi) - m^2\phi^*\phi - \lambda(\phi^*\phi)^2.$$

The covariant derivatives contain both the charged and the neutral vector fields:

$$D^\mu = (\partial^\mu - ig \cdot T_i \cdot W_i^\mu - ig' \cdot Y \cdot B^\mu).$$

A spontaneous symmetry breaking follows if $m^2 < 0$. By a suitable transformation it is always possible to arrange the fields in such a way that the vacuum expectation value of the scalar field vanishes for the upper component:

$$\phi_0 = \langle 0| \phi |\rangle = \frac{1}{\sqrt{2}} \begin{pmatrix} 0 \\ v \end{pmatrix},$$

$$\Rightarrow \frac{1}{8} v^2 \cdot ((g' B_\mu - g A_\mu^3)^2 + g^2 (A_\mu^1)^2 + g^2 (A_\mu^2)^2).$$

We redefine the fields of the charged gauge bosons:

$$W_\mu^\pm = (A_\mu^1 \mp A_\mu^2)/\sqrt{2}.$$

The masses are proportional to the vacuum expectation value:

$$M_W = (g/2) \cdot v.$$

The mass eigenstates of the neutral gauge bosons are the photon A and the neutral weak boson Z:

$$Z_\mu = \frac{-g A_\mu^3 + g' B_\mu}{\sqrt{g^2 + g'^2}}, \quad M_Z = v\sqrt{g^2 + g'^2}/2,$$

$$A_\mu = \frac{g B_\mu + g' A_\mu^3}{\sqrt{g^2 + g'^2}}, \quad M_A = 0.$$

It is useful to define a weak mixing angle:

$$\tan \theta_W = \frac{g'}{g},$$

$$A = \cos \theta_W \cdot B + \sin \theta_W \cdot A^3,$$

$$Z = -\sin \theta_W \cdot B + \cos \theta_W \cdot A^3.$$

This angle describes in particular the ratio of the masses of the charged and neutral weak bosons:

$$M_W/M_Z = g/\sqrt{g^2 + g'^2} = \cos\theta_W,$$
$$\sin^2\theta_W = 1 - \frac{M_W^2}{M_Z^2}.$$

The gauge bosons interact with the currents of the leptons:

$$L^w \sim \frac{g}{\sqrt{2}}(\bar{\nu}\gamma^\mu e_L^-) \cdot W_\mu^+ + h.c.$$

If the energies are much smaller than the masses of the weak bosons, the interaction can be described by an effective current–current interaction:

$$L^{\mathit{eff}} = \frac{g^2}{2M_W^2} \cdot (\bar{\nu}\gamma^\mu e_L) \cdot (\bar{e}\gamma_\mu \nu_L) = \frac{4G_F}{\sqrt{2}} \cdot (\bar{\nu}\gamma^\mu e_L) \cdot (\bar{e}\gamma_\mu \nu_L).$$

Thus the Fermi constant is given by the gauge coupling constant and the mass of the weak boson:

$$\frac{G_F}{\sqrt{2}} = \frac{g^2}{8M_W^2}.$$

The vacuum expectation value is directly related to the Fermi constant, which is measured:

$$v^2 = \frac{1}{\sqrt{2}G_F},$$
$$v \cong 246\,\text{GeV}.$$

The electromagnetic coupling constant is given by the gauge coupling constant and the weak mixing angle:

$$e = g \cdot \sin\theta_W = g' \cdot \cos\theta_w = gg'/\sqrt{g^2 + g'^2}.$$

The interaction of the neutral weak boson with the leptons defines the weak neutral current:

$$L^Z = \frac{g}{\cos\theta_W} Z^\mu j_\mu^n.$$

The neutral current is a superposition of the neutral isospin current and the electric current:

$$j_\mu^n = j_\mu^3 - \sin^2\theta_W j_\mu^e,$$

$$j_\mu^3 = \frac{1}{2}\left(\bar{\nu}\gamma_\mu\nu - e\gamma_\mu e\right)_L,$$

$$j_\mu^e = -\bar{e}\gamma_\mu e.$$

For the effective current–current interaction we find:

$$H^w = \frac{4G_F}{\sqrt{2}}\left(j_\mu^+ \cdot j_-^\mu + j_\mu^n \cdot j_n^\mu\right).$$

As mentioned before, the neutral current interaction violates parity and it was observed in 1968 at the Stanford Linear Accelerator Center in the scattering of electrons and protons.

Using the Fermi constant, the SU(2) coupling constant and the weak mixing angle, one can calculate the masses of the weak bosons:

$$\frac{G_F}{\sqrt{2}} = \frac{g^2}{8M_W^2},$$

$$M_W = \left(\frac{\pi\alpha}{\sqrt{2}G}\right)^{1/2}\frac{1}{\sin\theta_W} \cong \frac{37.3}{\sin\theta_W}\,\text{GeV},$$

$$M_Z = \frac{M_W}{\cos\theta_W} \cong \frac{74.6}{\sin 2\theta_W}\,\text{GeV}.$$

The weak bosons were finally discovered in 1983 at CERN. They were produced in proton–antiproton collisions. The decays of the weak bosons into leptons were observed:

$$W^+ \to \mu^+\nu_\mu, \quad Z \to \mu^+\mu^-.$$

The charged weak boson can also decay into a positron and a neutrino. The width for this decay is given by the Fermi constant and the mass of the weak boson:

$$\Gamma(W^+ \to e^+\nu_e) = \frac{g^2 M_W}{48\pi} = \frac{\alpha \cdot M_W}{12\sin^2\theta_W} = \frac{G_F \cdot M_W^3}{6\sqrt{2}\pi}.$$

Today the masses of the weak bosons and the weak mixing angle are rather well determined:

$$\begin{aligned} M_W &\cong 80.39\,\mathrm{GeV}, \\ M_Z &\cong 91.19\,\mathrm{GeV}, \\ \sin^2\theta_W &\cong 0.2312, \\ \theta_W &\cong 28.74^\circ. \end{aligned}$$

The electron mass is also generated by the spontaneous symmetry breaking. It is given by the vacuum expectation value of the scalar field and a Yukawa coupling constant G. This coupling constant cannot be calculated, thus the electron mass remains a free parameter:

$$\begin{aligned} L^m &= -G \cdot \bar{E}_L \cdot \phi \cdot E_R + h.c., \\ G &= \sqrt{2}\left(\frac{m_e}{v}\right) \cong 2,94 \cdot 10^{-6}. \end{aligned}$$

Two Leptons and Quarks

Now we consider the electroweak gauge theory of the electron, the electron neutrino, the up quark and the down quark. Both the leptons and the quarks form SU(2) doublets:

$$\begin{pmatrix} \nu_e \\ e^- \end{pmatrix}_L, \quad \begin{pmatrix} u \\ d \end{pmatrix}_L.$$

Since the effective mass of the down quark is slightly larger than the mass of the up quark, the neutron mass is larger than the proton mass. The neutron decays into the proton after emitting an electron and an antineutrino:

$$n \to p + e^- + \bar{\nu}_e.$$

This beta decay can be described by the current–current interaction and the Fermi constant:

$$H^W = \frac{4G_F}{\sqrt{2}}(J_\mu^* \cdot J^\mu).$$

The weak current is the sum of the left-handed lepton current and the left-handed quark current:

$$J_\mu = \bar{\nu}\gamma_\mu e_L^- + \bar{u}\gamma_\mu d_L.$$

The two leptons and the two quarks are considered to be a quark–lepton family:

$$\begin{pmatrix} \nu_e & u \\ e^- & d \end{pmatrix}.$$

Here are the hypercharges of the quarks:

$$Y(u_L) = Y(d_L) = 1/3,$$
$$Y(u_R) = 4/3, \quad Y(d_R) = -2/3.$$

The left-handed fermions form SU(2) doublets, the right-handed fermions SU(2) singlets:

$$E_L = \begin{pmatrix} \nu_e \\ e^- \end{pmatrix}_L, \quad e_R^-,$$

$$Q_L = \begin{pmatrix} u \\ d \end{pmatrix}_L, \quad (u)_R, \quad (d)_R.$$

The Lagrange density of the fermions is the sum of an electron term and a quark term:

$$L^f = L^e + L^q,$$

$$L^e = \bar{E}_L \left(i\not{\partial} + \frac{g}{2}\tau_i \not{W}_i - \frac{g'}{2}\not{B} \right) E_L + \bar{e}_R \left(i\not{\partial} - g'\not{B} \right) e_R,$$

$$L^q = \bar{Q}_L \left(i\not{\partial} + \frac{g}{2}\tau_i \not{A}_i + \frac{g'}{6}\not{B} \right) Q_L + \bar{u}_R \left(i\not{\partial} + \frac{2}{3}g'\not{B} \right) u_R$$
$$+ \bar{d}_R \left(i\not{\partial} - \frac{1}{3}g'\not{B} \right) d_R.$$

The neutral current is the superposition of the neutral isospin current and the electric current:

$$j_\mu^n = j_\mu^3 - \sin^2\theta_w j_\mu^e,$$

$$j_\mu^3 = \frac{1}{2}(\bar{\nu}\gamma_\mu\nu - e\gamma_\mu e)_L + \frac{1}{2}(\bar{u}\gamma_\mu u - \bar{d}\gamma_\mu d)_L,$$

$$j_\mu^e = -\bar{e}\gamma_\mu e + \frac{2}{3}\bar{u}\gamma_\mu u - \frac{1}{3}\bar{d}\gamma_\mu d.$$

Four Leptons and Quarks

In 1936 a new charged lepton was discovered in the cosmic rays — the muon. It has a mass of about 106 MeV and decays into an electron, an electron antineutrino and a muon neutrino:

$$\mu^- \to \nu_\mu + (\bar{\nu}_e + e^-).$$

This decay is similar to the beta decay. The muon turns into a muon neutrino by emitting a weak boson, which decays into an electron and an electron antineutrino. The muon and its neutrino form another lepton–quark family together with the strange and the charm quark:

$$\begin{pmatrix} \nu_\mu & c \\ \mu^- & s \end{pmatrix}.$$

The charged weak current has now four terms:

$$J_\mu^- = \bar{\nu}_e\gamma_\mu e_L^- + \bar{u}\gamma_\mu d_L + \bar{\nu}_\mu\gamma_\mu\mu_L^- + \bar{c}\gamma_\mu s_L.$$

Since the mass of the charm quark is much larger than the mass of the strange quark, the charm quark will decay into a strange quark after emitting a virtual weak boson. One would expect that particles that consist of at least one strange quark are stable. However this is not the case. A strange quark can also decay. It emits a virtual weak boson and turns into an up quark. This decay is allowed since the quarks mix. The quarks, which interact via the weak interactions, are not mass eigenstates, but mixtures of mass eigenstates:

$$J_\mu^- = \bar{\nu}_e \gamma_\mu e_L^- + \bar{u}\gamma_\mu d'_L + \bar{\nu}_\mu \gamma_\mu \mu^- + \bar{c}\gamma_\mu s'_L,$$

$$d' = \cos\theta_c \cdot d + \sin\theta_c \cdot s,$$

$$s' = -\sin\theta_c \cdot d + \cos\theta_c \cdot s.$$

The mixing angle is called the Cabibbo angle, introduced in 1963 by Nicola Cabibbo. According to the experiments this angle is about 13 degrees:

$$\sin\theta_c = 0.2255 \pm 0.0019.$$

Until today it is not understood, why the quarks mix. Probably this problem is related to the problem of mass generation for the quarks and leptons.

The current–current interaction describes now besides the beta decay also the muon decay as well as the decay of the strange and of the charmed particles. The muon decay is given by the current–current interaction:

$$H^W = \frac{G_F}{\sqrt{2}} \bar{\nu}_\mu \gamma_\mu (1-\gamma_5)\mu^- \cdot e^- \gamma_\mu (1-\gamma_5)\nu_e.$$

Using this interaction, one can calculate the lifetime of the muon (the mass of the electron is neglected):

$$\tau_\mu^{-1} = \frac{G_F^2 m_\mu^5}{192\pi^3}.$$

The lifetime of the muon is very well measured:

$$\tau_\mu \cong 2.197019 \times 10^{-6}\,\mathrm{s}.$$

Using this value, one can calculate the Fermi constant:

$$G_F \approx 1.166 \times 10^{-5}\,\mathrm{GeV}^{-2}.$$

The mixing of the quarks appears only in the charged weak current, not in the neutral current. Here the mixing terms cancel:

$$J_\mu^n = J_\mu^3 - \sin^2\theta_w J_\mu^e,$$

$$J_\mu^3 = \frac{1}{2}(\bar{\nu}_e\gamma_\mu\nu_e - \bar{e}\gamma_\mu e)_L + \frac{1}{2}(\bar{u}\gamma_\mu u - \bar{d}\gamma_\mu d)_L$$

$$+\frac{1}{2}(\bar{\nu}_\mu\gamma_\mu\nu_\mu - \bar{\mu}\gamma_\mu\mu)_L + \frac{1}{2}(\bar{c}\gamma_\mu c - \bar{s}\gamma_\mu s)_L,$$

$$J_\mu^e = -\bar{e}\gamma_\mu e - \bar{\mu}\gamma_\mu\mu + \frac{2}{3}\bar{u}\gamma_\mu u - \frac{1}{3}\bar{d}\gamma_\mu d + \frac{2}{3}\bar{c}\gamma_\mu c - \frac{1}{3}\bar{s}\gamma_\mu s.$$

The mixing terms cancel only if there is a charm quark besides the strange quark, otherwise there would be a term in the neutral current connecting the down quark and the strange quark. Such a term is excluded since it would have been observed in the decay of a neutral K meson into two muons.

Six Leptons and Quarks

The third charged lepton, the tau lepton (mass $\sim$ 1777 MeV) was discovered in 1975 at the Stanford Linear Accelerator Center. Again this new lepton has its own neutrino, the tau neutrino. Now there are three doublets of leptons:

$$\begin{pmatrix} \nu_e & \nu_\mu & \nu_\tau \\ e & \mu & \tau \end{pmatrix}.$$

The tau lepton decays by emitting a virtual weak boson and becoming a tau neutrino. Here are examples of possible decays:

$$\tau^- \to \mu^- + \bar{\nu}_\mu + \nu_\tau,$$

$$\tau^- \to \nu_\tau + \pi^-,$$

$$\tau^- \to \nu_\tau + \pi^- + \pi^0.$$

The tau lepton and the tau neutrino belong to the third lepton–quark family, which consists of the top quark, the bottom quark, the tau lepton and its neutrino:

$$\begin{pmatrix} \nu_\tau & t \\ \tau^- & b \end{pmatrix}.$$

The weak decay of a b quark proceeds via the quark mixing. Most of the time a b quark decays into a charm quark, but sometimes it also decays into an up quark.

In the case of four quarks there is only one mixing angle — the Cabibbo angle. But for six quarks one has three mixing angles:

$$\begin{pmatrix} u & c & t \\ d' & s' & b' \end{pmatrix}.$$

The mixing of the quarks is described by a unitary matrix — the CKM matrix (Nicola Cabibbo, Makoto Kobayashi, Toshihide Maskawa):

$$\begin{pmatrix} d' \\ s' \\ b' \end{pmatrix} = \begin{pmatrix} V_{ud} & V_{us} & V_{ub} \\ V_{cd} & V_{cs} & V_{cb} \\ V_{td} & V_{ts} & V_{tb} \end{pmatrix} \begin{pmatrix} d \\ s \\ b \end{pmatrix}.$$

Here are the average values of the matrix elements, determined by many experiments:

$$|V_{CKM}| \approx \begin{pmatrix} 0.974 & 0.225 & 0.004 \\ 0.225 & 0.973 & 0.041 \\ 0.009 & 0.040 & 0.999 \end{pmatrix}.$$

In the case of four quarks the mixing matrix is real, but for six quarks the matrix elements are complex numbers. The CKM matrix is described not only by three mixing angles, but also one phase parameter. There are various ways to describe the matrix. We use the following possibility given by the product of three simple rotation matrices:

$$V_{CKM} = \begin{pmatrix} c_u & s_u & 0 \\ -s_u & c_u & 0 \\ 0 & 0 & 1 \end{pmatrix} \cdot \begin{pmatrix} e^{-i\varphi} & 0 & 0 \\ 0 & c & s \\ 0 & -s & c \end{pmatrix} \cdot \begin{pmatrix} c_d & -s_d & 0 \\ s_d & c_d & 0 \\ 0 & 0 & 1 \end{pmatrix},$$

$$c_u = \cos\theta_u, \quad s_u = \sin\theta_u,$$
$$c = \cos\theta, \quad s = \sin\theta,$$
$$c_d = \cos\theta_d, \quad s_d = \sin\theta_d.$$

Here are the experimental values of these angles:

$$\theta_u \approx 5.0^\circ, \quad \theta_d \approx 13.0^\circ, \quad \theta \approx 2.4^\circ.$$

The phase parameter describes the CP violation, which was discovered in 1964 in the decay of K mesons. The experiments indicate that the phase parameter is close to 90 degrees.

The Masses of Neutrinos

In the Standard Model neutrinos are massless. They are described by left-handed fermions. But in 1998 neutrino oscillations were discovered. Since such oscillations can exist only, if the masses of at least two neutrinos are different from zero, there is a problem with the Standard Model. It must be extended to include right-handed neutrinos, which are needed for the mass terms.

The mass terms of the quarks generate the mixing of the quarks. If neutrinos have mass, then one would expect also a mixing of the leptons. If a weak boson interacts with an electron, which becomes a neutrino, this neutrino is not a mass eigenstate, but a mixing of three mass eigenstates. The mixing is again described by a unitary matrix, similar to the CKM matrix:

$$U = \begin{pmatrix} U_{e1} & U_{e2} & U_{e3} \\ U_{\mu 1} & U_{\mu 2} & U_{\mu 3} \\ U_{\tau 1} & U_{\tau 2} & U_{\tau 3} \end{pmatrix}.$$

This matrix is again described by three mixing angles and one phase parameter:

$$U = \begin{pmatrix} c_l & s_l & 0 \\ -s_l & c_l & 0 \\ 0 & 0 & 1 \end{pmatrix} \cdot \begin{pmatrix} e^{-i\varphi} & 0 & 0 \\ 0 & c & s \\ 0 & -s & c \end{pmatrix} \cdot \begin{pmatrix} c_\nu & -s_\nu & 0 \\ s_\nu & c_\nu & 0 \\ 0 & 0 & 1 \end{pmatrix},$$

$$c_l = \cos\theta_l, \quad s_l = \sin\theta_l,$$

$$c = \cos\theta, \quad s = \sin\theta,$$

$$c_\nu = \cos\theta_\nu, \quad s_\nu = \sin\theta_\nu.$$

The first angle is the *reactor angle*, the second angle is the *atmospheric angle*, the third angle is the *solar angle*. The experiments give the following values for the mixing angles and the mass differences:

$$\theta_l = \theta_{re} \approx 9^\circ,$$

$$38^\circ \leq \theta = \theta_{at} \leq 52^\circ,$$

$$31^\circ \leq \theta_\nu = \theta_{sun} \leq 37^\circ,$$

$$\Delta m_{21}^2 \approx 7.6 \times 10^{-5}\,\mathrm{eV}^2,$$

$$\Delta m_{32}^2 \approx 2.4 \times 10^{-3}\,\mathrm{eV}^2.$$

Nothing is known thus far about the phase parameter, which describes the CP violation for the leptons.

In the neutrino oscillations one can measure only the differences of the masses of the neutrinos. The absolute values of the neutrino masses are still unknown. If we assume that the first neutrino mass is zero, the masses of the other two neutrinos are fixed:

$$m_1 \approx 0.000\,\mathrm{eV},$$

$$m_2 \approx 0.009\,\mathrm{eV},$$

$$m_3 \approx 0.049\,\mathrm{eV}.$$

In this case the neutrino masses are much smaller than the masses of the charged leptons. Probably this is the case. But it is also possible that the three neutrino masses are very similar — in this case the neutrino masses could be in the region of about 1 eV.

A neutrino mass term could be similar to the electron mass term:

$$\nu = (\nu_L, \nu_R),$$

$$m(\bar{\nu}\nu) = m(\bar{\nu}_L \nu_R + \bar{\nu}_R \nu_L).$$

Since the neutrinos are neutral, the neutrino mass term could be different — it might be a Majorana mass term. Such a mass term, introduced in about 1938 by Ettore Majorana, describes the transition from a left-handed neutrino field to a right-handed antineutrino field:

$$m^\nu \propto m(\bar{\nu}_L \nu_R^c + \bar{\nu}_R^c \nu_L).$$

If such mass terms are present, there is no conserved lepton number. It is possible to measure a Majorana term by investigating the neutrino-less double beta decay. In a nucleus two neutrons decay and two antineutrinos are produced. They can annihilate via the Majorana mass term:

$$n + n \to p + e^- + (\bar{\nu}_e + \bar{\nu}_e) + e^- + p \to p + e^- + e^- + p.$$

The nucleus then decays by emitting two electrons, but no antineutrinos. For example, the element tellurium might decay via a neutrino-less double beta decay.

The neutrino-less double beta decay has not been observed. Since the decay amplitude is given by the neutrino mass, it is possible to find a limit for a Majorana neutrino mass: it must be smaller than 0.1 eV.

A neutrino mass could also have both Dirac terms and Majorana terms; in this case it can be written as follows:

$$L \sim \frac{1}{2}(\bar{\nu}_L, \bar{\nu}_L^c) \begin{pmatrix} m & D \\ D & M \end{pmatrix} \begin{pmatrix} \nu_R^c \\ \nu_R \end{pmatrix}.$$

Here D is the Dirac mass term, m and M are the two Majorana mass terms. If D is zero, the neutrino mass is a Majorana mass. If m and

M are zero, it is a Dirac mass. An interesting case is when $m = 0$ and $D \ll M$:

$$M_\nu = \begin{pmatrix} 0 & D \\ D & M \end{pmatrix}.$$

Then one has a very small and very large neutrino mass:

$$m_1 \approx \frac{D^2}{M}, \quad m_2 \approx M.$$

This is the "seesaw" mechanism. If M is large and D about equal to the mass of the charged lepton, the neutrino mass is much smaller than the mass of the charged lepton.

For example, we can calculate the mass of the tau neutrino, if the Dirac mass is given by the mass of the tau lepton and the large Majorana mass is taken to be 10^{10} GeV. One finds:

$$m_{\nu T} \approx 0.3\,\text{eV}.$$

The Higgs Particle in the Standard Model

In the Standard Model the mass of the Higgs boson is a function of the vacuum expectation value of the scalar field and the coupling constant:

$$M_H = \sqrt{2\lambda} \cdot v.$$

Since the coupling constant is unknown, the mass of the Higgs boson cannot be calculated, but some information about the mass can be obtained. The weak bosons interact with the Higgs boson. A weak boson can emit and reabsorb a virtual Higgs boson, and the masses of the weak bosons are changed slightly. Since the masses of the weak bosons are known quite well, one can obtain a limit on the mass of the Higgs boson — it must be smaller than 180 GeV.

Using the LEP accelerator at CERN, one could also obtain a lower limit. The mass of the Higgs boson must be larger than

115 GeV, otherwise it would have been observed in the LEP collisions. Thus we find the mass range:

$$115\,\mathrm{GeV} < M_H < 180\,\mathrm{GeV}.$$

In 2012 one has discovered with the new LHC accelerator at CERN a new neutral scalar boson with a mass of about 125 GeV. It decays e.g. into a real and a virtual weak boson or into two photons. This boson could be the Higgs boson. But it could also be an excitation of the neutral weak boson, if the weak bosons have an internal structure. Further experiments are needed.

Chapter 16

Grand Unification

In the Standard Model of particle physics there are 25 fundamental constants that cannot be calculated, but have to be determined by experiments: the three coupling constants of the electromagnetic, weak and strong interactions, the six masses of the leptons, the six masses of the quarks, the eight mixing parameters for the flavor mixing of the quarks and leptons, the mass of the weak boson and the mass of the Higgs boson.

Furthermore in the Standard Model one does not understand why there are three families of leptons and quarks and why there are three colors of the quarks. Why is the electric charge of the proton equal to the electric charge of the positron? The electric charge of the down quark is $-1/3\ e$, the electric charge of the electron is $-e$. The ratio 3 of the charges is probably related to the fact that the quarks have three colors.

An interesting way to understand the charge quantization is an embedding of the gauge group of the Standard Model into a larger group. Let us consider the first family consisting of eight fermions — three doublets of the colored up and down quarks and one doublet of leptons, the electron and its neutrino:

$$\begin{pmatrix} \nu_e & u_r & u_g & u_b \\ e & d_r & d_g & d_b \end{pmatrix}.$$

The lepton number can be interpreted as a fourth color. The color group SU(3) would then be a subgroup of the group SU(4). Of course, this SU(4) symmetry could not be exact, since in this case there would be no difference between leptons and quarks.

After the symmetry breaking one has the color group SU(3) and a group U(1):

$$\mathrm{SU}(4) \Rightarrow \mathrm{SU}(3) \otimes \mathrm{U}(1).$$

The observed charged weak bosons interact only with left-handed fermions. But perhaps at very high energy there are also very massive weak bosons, interacting with the right-handed fermions. In this case the group SU(2) of the weak isospin would be a subgroup of a larger group:

$$\mathrm{SU}(2)_L \subset \mathrm{SU}(2)_L \otimes \mathrm{SU}(2)_R.$$

The violation of parity in the weak interactions would be a consequence of the mass difference between the masses of the gauge bosons. The masses of the gauge bosons that couple to the right-handed fermions must be larger than 1 TeV, otherwise they would have been observed in the experiments.

The leptons and quarks are now described by left-handed and right-handed doublets:

$$\begin{pmatrix} \nu_e & u \\ e^- & d \end{pmatrix}_L \quad \begin{pmatrix} \nu_e & u \\ e^- & d \end{pmatrix}_R$$

The right-handed neutrino is new — it is not present in the Standard Model. The gauge group of the Standard Model is a subgroup of a larger group G:

$$\mathrm{SU}(3) \otimes \mathrm{SU}(2)_L \otimes \mathrm{U}(1) \subset \mathrm{SU}(4) \otimes \mathrm{SU}(2)_L \otimes \mathrm{SU}(2)_R = G.$$

In a gauge theory, based on the group G, there are besides the eight gluons seven color triplet gauge bosons, which induce transitions between leptons and quarks. Thus the proton would not be stable, but will decay, e.g. into three neutrinos and a charged pion. Since thus far no proton decay has been observed, the masses of these gauge bosons must be larger than 1 TeV.

The breaking of the G-symmetry can be arranged in two steps. First the masses of the seven SU(4) gauge bosons are generated, then the masses of the gauge bosons of the group $SU(2)_R$. This mass generation can also be arranged in one step:

$$G = SU(4) \otimes SU(2)_L \otimes SU(2)_R \Rightarrow SU(3)_c \otimes SU(2)_L \otimes U(1).$$

In the Standard Model the photon is a mixture, since the electric charge is a superposition of the neutral isospin charge and the hypercharge. In the group G the electric charge is a superposition of three charges:

$$Q = \frac{1}{2}(B - L) + T_{3L} + T_{3R},$$

$$Q = \frac{1}{2}\begin{pmatrix} 1/3 & 0 & 0 & 0 \\ 0 & 1/3 & 0 & 0 \\ 0 & 0 & 1/3 & 0 \\ 0 & 0 & 0 & -1 \end{pmatrix}$$

$$+ \begin{pmatrix} +1/2 & 0 \\ 0 & -1/2 \end{pmatrix}_L + \begin{pmatrix} +1/2 & 0 \\ 0 & -1/2 \end{pmatrix}_R .$$

In a gauge theory, based on the group G, the electric charges are not free parameters. Since there are three colors, the charge of the down quark is $-1/3$ e. Thus the electric charges are quantized!

The coupling constants of QCD and of the SU(2) gauge theory decrease with increasing energy ("asymptotic freedom"), but the coupling constant of the U(1) gauge theory increases. It might be that at very high energy, e.g. at about 10^{15} GeV, the three coupling constants converge (Fig. 16.1). At this energy the electroweak interactions and QCD would be unified, if the color group and the gauge group of the electroweak interactions are subgroups of a simple group, e.g. the group SU(5) or the group SO(10).

The Group SU(5)

The rank of the Standard Model is 4, the rank of the group SU(5) is also four, thus this group is the smallest gauge group for a unified theory. The coupling constants of the Standard Model converge at

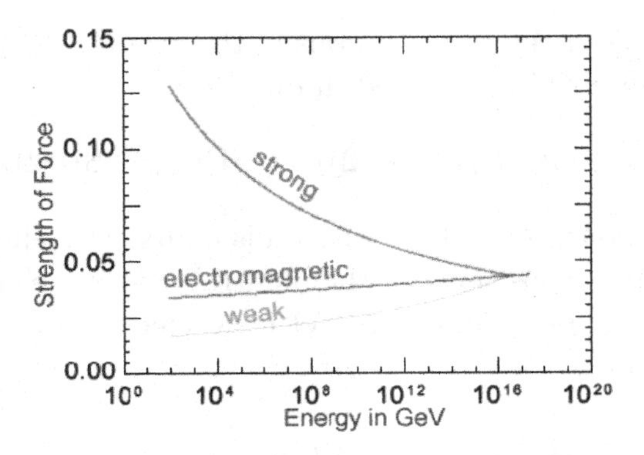

Fig. 16.1. Convergence of the coupling constants.

very high energies:

$$\mathrm{SU}(3) \otimes \mathrm{SU}(2) \otimes \mathrm{U}(1) \Rightarrow \mathrm{SU}(5).$$

The group SU(5) has 24 generators and is described by five-dimensional Hermitian matrices. Thus there are 24 gauge bosons, among them the eight gluons and the four gauge bosons of the electroweak theory. The other twelve gauge bosons will be discussed below.

The first family of the Standard Model consists of 15 left-handed fermions: three u quarks, three d quarks, the corresponding anti-quarks, the electron, the positron and the electron neutrino. In the SU(5) theory these 15 fermions are described by a five-dimensional and a ten-dimensional representation. The five-dimensional representation describes the three down antiquarks, the electron and its neutrino:

$$(\bar{5}) = \begin{pmatrix} \bar{d}_r \\ \bar{d}_g \\ \bar{d}_b \\ \nu \\ e \end{pmatrix}.$$

The 10-representation contains the three up quarks and the corresponding antiquarks, the three down quarks and the positron:

$$(10) = \frac{1}{\sqrt{2}} \begin{pmatrix} 0 & \bar{u}_b & -\bar{u}_g & -u_r & -d_r \\ -\bar{u}_b & 0 & \bar{u}_r & -u_g & -d_g \\ \bar{u}_g & -\bar{u}_r & 0 & -u_b & -d_b \\ u_r & u_g & u_b & 0 & -e^+ \\ d_r & d_g & d_b & e^+ & 0 \end{pmatrix}.$$

Thus the leptons and quarks of one family are described by the sum of two representations:

$$\bar{5} \oplus 10 \Rightarrow (3,2) \oplus 2(\bar{3},1) \oplus (1,2) \oplus (1,1).$$

As in the group G the electric charges are quantized. The sum of the charges in each representation must vanish, thus the charges of the quarks and the number of colors are correlated:

$$3Q(\bar{d}) + Q(e^-) = 0.$$

The SU(5) theory has 24 gauge bosons. Twelve of these gauge bosons are the gauge bosons of the Standard Model. The other twelve gauge bosons, the (X-Y)-bosons, generate transitions between leptons and quarks:

$$24 = (8,1,1) \oplus (1,3,1) \oplus (1,1,1) \oplus (3,2,1) \oplus (3^*,2,1),$$

$$\begin{pmatrix} X \\ Y \end{pmatrix} = (3,2,1)$$

The (X-Y)-bosons are also called "lepto-quark bosons," since they couple to leptons and quarks. They must have a very high mass, since two quarks inside the proton can produce a virtual lepto-quark boson, which decays into a positron and an antiquark. Thus the proton can decay into a neutral pion and a positron:

$$u + d \to Y \to \bar{u} + e^+,$$

$$p \to \pi^0 + e^+.$$

Since this decay has not been observed, one obtains a limit for the masses of the lepto-quark bosons — they must be larger than 10^{14} GeV.

The SU(5) symmetry is broken in two steps. In the first step the masses of the lepto-quark bosons are generated. The group SU(5) breaks down to the gauge group of the Standard Model. This breaking can be arranged by an adjoint representation of scalar fields:

$$\mathrm{SU}(5) \Rightarrow \mathrm{SU}(3) \otimes \mathrm{SU}(2) \otimes \mathrm{U}(1).$$

The masses of the weak bosons are generated by scalar fields transforming as a fundamental representation of SU(5) — a five-dimensional representation. This representation contains the doublet, which is needed for the symmetry breaking of the electroweak gauge group. But this representation contains also three colored fields, which can generate the proton decay. Their masses must be very large, at least 10^{14} GeV. Thus in the five-dimensional representation, two different mass scales appear. The ratio of these mass scales is less than 10^{-12} — this is the *hierarchy problem*.

The electroweak mixing angle is given by the ratio of the two coupling constants in the electroweak theory, which has been determined by the experiments:

$$\tan\theta_w = g'/g \cong 0.55.$$

The normalization of the weak isospin generators and of the hypercharge is determined by the group SU(5):

$$\mathrm{tr}T_3^2 = \frac{1}{2}, \quad \mathrm{tr}(Y/2)^2 = \frac{5}{6}.$$

Thus the electroweak mixing angle at the energy, where the interactions are unified, can be calculated:

$$\tan\theta_w(M_X) = g'/g = \sqrt{3/5} \approx 0.77,$$
$$\sin^2\theta_w(M_X) = 3/8,$$
$$\theta_w(M_X) \cong 37.8^\circ.$$

With the LEP accelerator at CERN one has measured the fine-structure constant and the electroweak mixing angle at the mass of the Z boson with great precision:

$$\alpha(M_Z^2)^{-1} = 128.93 \pm 0.02,$$
$$\sin^2\theta_w(M_Z) = 0.23102 \pm 0.00005.$$

For a comparison we have to extrapolate the electroweak mixing angle from the Z mass to the unification energy:

$$\sin^2\theta_w(M_X) = \frac{1}{6} + \frac{5}{24} \cdot \left(1 - \frac{11 \cdot \alpha\left(M_Z^2\right)}{2\pi} \ln \frac{M_X^2}{M_Z^2}\right).$$

Using those values, we can estimate the energy where the electroweak and the strong interactions are unified:

$$M_X \approx 10^{13}\ \text{GeV}.$$

The mass of the X bosons determines the lifetime of the proton, about 10^{31} years. But according to the experiments the lifetime of the proton is longer than 10^{33} years. Thus the mass of the X bosons cannot be less than 10^{15} GeV — this is a problem for the SU(5) theory.

If we extrapolate the QCD coupling from the energy of unification to the mass of the Z boson, we find:

$$\alpha_s(M_Z^2) \approx 0.07.$$

But the experiments give a larger value for this coupling constant — another problem for the SU(5) theory:

$$\alpha_s(M_Z^2) \approx 0.12.$$

Thus there are two problems — the SU(5) theory cannot be the correct theory of the unification of quantum chromodynamics and the electroweak theory.

The Group SO(10)

Some unitary groups are isomorphic to orthogonal groups. The group SU(2) is isomorphic to SO(3), and the product SU(2) × SU(2) is isomorphic to SO(4). The group SU(4) is isomorphic to SO(6). Thus the product SO(4) × SO(6) is isomorphic to the product SU(4) × SU(2) × SU(2), which is the group G that we discussed above. This group is a subgroup of the group SO(10):

$$\mathrm{SU}(4) \otimes \mathrm{SU}(2)_L \otimes \mathrm{SU}(2)_R \cong \mathrm{SO}(6) \otimes \mathrm{SO}(4) \subset \mathrm{SO}(10).$$

The group SO(10) could be used to describe the unification of QCD and the electroweak theory. It describes rotations in a 10-dimensional space. The simplest representation of SO(10) is the (10) representation, which is analogous to the vector representation of the rotation group SO(3).

The group SO(3) has a two-dimensional spinor representation — analogously the group SO(10) has a 16-dimensional spinor representation. Each of the three families of leptons and quarks can be described by a (16) representation, e.g.:

$$\begin{pmatrix} \nu_e & u & u & u & \bar{u} & \bar{u} & \bar{u} & \bar{\nu}_e \\ e^- & d & d & d & \bar{d} & \bar{d} & \bar{d} & e^+ \end{pmatrix}.$$

In the Standard Model the first family has only 15 left-handed fermions — there is no left-handed antineutrino. The group SO(10) requires the existence of another antineutrino.

The group SU(5) is a subgroup of SO(10):

$$\mathrm{SO}(10) \supset \mathrm{SU}(5) \otimes \mathrm{U}(1).$$

The spinor representation of SO(10) is a sum of three representations of SU(5). The left-handed singlet is the new antineutrino:

$$16 = \bar{5} \oplus 10 \oplus 1.$$

The breaking of SO(10) to the gauge group of the Standard Model via SU(5) implies that the renormalization of the three coupling constants is the same as in the SU(5) theory — thus one has the same problem. However a different symmetry breaking is interesting:

$$\mathrm{SO}(10) \Rightarrow \mathrm{SO}(6) \otimes \mathrm{SO}(4) = G \Rightarrow \mathrm{SU}(3) \otimes \mathrm{SU}(2) \otimes \mathrm{U}(1),$$

$$(16) \Rightarrow (4, 2, 1) \oplus (\bar{4}, 1, 2).$$

In the SU(5) theory the neutrinos cannot have a mass, but in SO(10) the neutrinos should have a mass, since there are left-handed and right-handed neutrinos. The neutrino mass might be a Dirac mass, a Majorana mass or a mass due to the seesaw mechanism.

The breaking of SO(10) can be arranged as follows:

I: Breaking SO(10) → G

$$\mathrm{SO}(10) \Rightarrow \mathrm{SU}(4) \otimes \mathrm{SU}(2)_L \otimes \mathrm{SU}(2)_R = G$$

II: Breaking G → gauge group of Standard Model

$$G \Rightarrow \mathrm{SU}(3) \otimes \mathrm{SU}(2)_L \otimes \mathrm{U}(1)$$

III: Breaking of gauge group of Standard Model

$$\mathrm{SU}(3) \otimes \mathrm{SU}(2)_L \otimes \mathrm{U}(1) \Rightarrow \mathrm{SU}(3) \otimes \mathrm{U}(1)$$

The first breaking can be arranged with a (54) representation of scalar fields. The group G can be broken with a ((54) + (16)) representation. The third breaking can be arranged with a (10) representation of scalars.

In the SU(5) theory there are two steps of symmetry breaking, in SO(10) one more step is necessary. In this case three different energy scales are involved:

(1) The energy of Grand Unification.
(2) The energy of the breaking of the group G.
(3) The energy of electroweak symmetry breaking.

Since there are three different energy scales, it is not possible to calculate the energy scale of the Grand Unification, using the observed coupling constants, as was done in the SU(5) theory. The energy of the Grand Unification depends also on the energy of the breaking of G, in particular on the masses of the gauge bosons of the group $\mathrm{SU}(2)_R$. These masses are unknown, but they must be larger than about 3 TeV. The unification of the various coupling constants would be in the range 10^{15}–10^{16} GeV — it is about two orders of magnitude higher than in the SU(5) theory.

The lifetime of the proton is expected to be in the range 10^{34}–10^{36} years. The proton decay has not been observed, but it might be observed in the future. If the proton has a lifetime longer than 10^{35} years, the decay cannot be seen, due to the background from neutrino–nucleus reactions.

Thus far there are no serious problems with the SO(10) group as the gauge group of Grand Unification. It might be the correct theory to describe the unification of the strong, electromagnetic and weak interactions.

Supersymmetry and Strings

In the Standard Model there are fermions and gauge bosons. For example, in QCD there are quarks, which are color triplets, and gluons, which are color octets. Many theorists speculate that at high energies a new symmetry appears — the *supersymmetry*. If this symmetry were unbroken, each fundamental particle would have a supersymmetric partner with a different spin. The partner of the electron would have spin zero — the *selectron*. For each quark there is a scalar *squark*. The partner of the photon is a *photino* with spin 1/2, and the partners of the gluons are the *gluinos* with spin 1/2.

Thus far no supersymmetric partner of the particles of the Standard Model have been observed. If they exist, their masses must be larger than 1 TeV.

There are attempts to generalize the quantum field theory to a "string theory." In such a theory the fundamental particles are not point-like, but small one-dimensional "strings." In particular, one tries to construct with the strings a theory of quantum gravity. The typical length of a string should be as small as the Planck length, which is given by the gravitational constant and the Planck constant:

$$l_P = \sqrt{\frac{\hbar G}{c^3}} \approx 1.62 \times 10^{-35}\text{m}.$$

But it remains unclear how a quantization of gravity can be arranged. According to the theory of General Relativity, the gravitational force is not a real force, but follows from the curvature of space-time. A quantization of gravity would imply that space and time must also be quantized, but nobody knows how to arrange this. Thus quantum field theory remains an interesting area of research in the future.

Appendix

Basic Constants

Speed of light:

$$c = 2.998 \times 10^{10} \text{ cm/s}$$

Quantum of action:

$$\hbar = 6.582 \times 10^{-22} \text{ MeV} \cdot \text{s}$$

Fine-structure constant:

$$\alpha = \frac{e^2}{4\pi\hbar c} \cong \frac{1}{137.04} \cong 0.00730.$$

Fermi constant:

$$\frac{G_F}{(\hbar c)} = 1.166 \times 10^{-5} \text{ GeV}^{-2}$$

Natural Units

In quantum physics it is useful to use natural units. The action is given as multiples of the Planck constant, while the velocity of light is the unit for velocities. Both constants are set to one:

$$\hbar = 6.582 \times 10^{-22} \text{ MeV} \cdot \text{s} \Rightarrow 1,$$

$$c = 299,792 \text{ km/s} \Rightarrow 1.$$

Relativity

Metric tensor:

$$g_{\mu\nu} = g^{\mu\nu} = \begin{pmatrix} 1 & 0 & 0 & 0 \\ 0 & -1 & 0 & 0 \\ 0 & 0 & -1 & 0 \\ 0 & 0 & 0 & -1 \end{pmatrix}.$$

Contravariant four-vectors:

$$x^\mu = (x^0, \vec{x}).$$

Covariant four-vectors:

$$x_\mu = g_{\mu\nu} x^\nu = (x^0, -\vec{x}).$$

$$p \cdot x = g_{\mu\nu} p^\mu x^\nu = p^0 x^0 - \vec{p}\vec{x}.$$

Four-momentum of particle with mass m:

$$p^2 = p^\mu p_\mu = E^2 - |\vec{p}|^2 = m^2.$$

Bibliography

J. Bjorken, S. Drell, *Relativistic Quantum Fields.*
McGraw-Hill, New York, 1965.

N. Bogoliubov, S. Shirkov, *Quantum Fields.*
Benjamin-Cummings, 1982.

L. Brown, *Quantum Field Theory.*
Cambridge University Press, 1992.

R. Feynman, *QED: The Strange Theory of Light and Matter.*
Princeton University Press, 1985.

C. Itzykson, J. Zuber, *Quantum Field Theory.*
McGraw-Hill, 1980.

M. Kaku, *Quantum Field Theory: A Modern Introduction.*
Oxford University Press, 1993

D. McMahon, *Quantum Field Theory Demystified.*
McGraw-Hill, 2008.

F. Mandl, G. Shaw, *Quantum Field Theory,*
John Wiley & Sons, Chichester, 1984.

M. Peskin, D. Schroeder, *An Introduction to Quantum Field Theory.*
Westview Press, 1995.

C. Quigg, *Gauge Theories of the Strong, Weak and Electromagnetic Interactions.*
Benjamin/Cummings, 1983.

P. Ramond, *Field Theory: A Modern Primer*,
Cambridge University Press, 1989.

G. Sterman, *Introduction to Quantum Field Theory*,
Cambridge University Press, 1993.

S. Weinberg, *The Quantum Theory of Fields*,
Cambridge University Press, 1995.

A. Zee, *Quantum Field Theory in a Nutshell*,
Princeton University Press, 2003.